SPECIAL ELECTRICAL MACHINES

SIMMI P. BURMAN

B.E. (Electrical Engineering)
Jabalpur (Govt.) Engineering College, Jabalpur
M.Tech. (Persuing), GBTU Lucknow
Power System and Energy

S.K. KATARIA & SONS®

Publisher of Engineering Books
4885/109, Prakash Mahal, Dr. Subhash Bhargav Lane,
Opposite Delhi Medical Association, Daryaganj, New Delhi–110002 (India)
Phone: +91-11-43551243; Mobile: +91-9871775858, 7678425209
e-mail: katariabook@gmail.com, katariabooks@yahoo.com
Website: www.skkatariaandsons.com
(e-books, subscriptions and bulk buying also available on our website)

Published by:

S.K. Kataria & Sons®

4885/109, Prakash Mahal, Dr. Subhash Bhargav Lane,
Opposite Delhi Medical Association, Daryaganj, New Delhi-110002 (India)
Phone: +91-11-43551243; Mobile: +91-9871775858, 7678425209
e-mail: katariabook@gmail.com, katariabooks@yahoo.com

First Edition: 2012; Second Edition: 2013; Reprint: 2026

Laser Type Setting by:
SARA ASSIGNMENTS, Delhi–110092.

Printed at :
A.S. OFFSET, Delhi

P Preface

The primary aim of presenting this book is to meet the needs of B.Tech (Electrical Engineering and Electrical and Electronics Engineering) students of UP Technical University for the subject of Special Electrical Machines. The subject matter has been presented in an easy comprehensive manner, logical sequence using simple language and to provide a detailed exposition of the fundamental.

The book consists of 5 divided units providing a comprehensive coverage of courses prescribed by UP Technical University for the subject of Special Electrical Machines. A large number of worked out examples, theoretical exercises and numerical problems, all of Degree standard, have been included in the text. The solution to illustrative examples covering almost all topics makes the entire presentation easy to follow.

Every care has been taken to eliminate misprints, omissions and errors but it is too much to expect that no in accuracy misprint/error has crept in and the author would be grateful to the readers for bringing to his notice any such errors/misprints/omissions, they may come across while going through the book.

The author shall feel satisfied that, if the book meets the needs of the students for whom it is meant. Suggestions and criticism for the improvement of the book shall be welcomed.

The author will like to thank all the scientists, Engineers. Authors and publishers, whose works and text have been the source of enlightenment, inspiration and guidance to me in the presentation.

Lastly, but not the least, the author is thankful to her publisher for bringing out the book in such a short time and for pricing it moderately deposit heavy cost of paper in printing.

January, 2013 **Simmi P. Burman**

A Acknowledgement

The author expresses her kind gratitude to, Dr. Raghuvir Kumar, Director of Babu Banarasi Das National Institute of Technology and Management and, Dr. Seethalekshmi. K, Director of BBDNITM, HoD EE Dept, Lucknow.

The author places on record his heart felt thanks to Mr. Bipin Singh, Senior Lecturer, BBDNITM for the encouragement extended to her.

The author also wishes to thank Mr. Jegan Raj. S, Senior Lecturer, Department of Electrical Engineering, BBDNITM and her colleagues for their consistent co-operation directly or indirectly in bringing out this book.

The author likes to remember the motivation initiated by her husband Mr. Avnish Burman, Deputy Manager, Hindustan Aeronautics Limited.

Simmi P Burman

To My Loving Son
SAUBHAGYA BURMAN
(JOY)

Contents

Chapter

1

Poly-phase AC Machines

1.1 INTRODUCTION

The induction machine is an important clam of electric machines, which finds wide applicability as a motor in industry and in its single-phase form in several domestic application. More than 85% of industrial motors in use today are infact induction motors.

It is substantially a constant-speed motor with a shunt characteristic, a few percent speed drop from no load to full load.

The torque developed in this motor has its origin in current induction in the rotor which is only possible at non-synchronous speed, hence the name asynchronous machine. On the other hand the torque in a synchronous machine is developed only at synchronous speed when the locking of the two field takes place. Therefore, the induction motor is not plagued by the stability problem inherent in the synchronous motor. As a consequence it inherently has a power factor less then unity which usually must be corrected by means of shunt capacitors at motor terminals. There is no simple and in expensive method of controlling the induction motor speed as is possible in a dc shunt motor.

A wide range of speed control is possible only by expensive circuit using (SCRs).

1.2 CONSTRUCTION

The stator of an induction motor is similar to that of a synchronous machine and is wound for three phases. Two types of construction are employed for the rotor-wound-rotor and squirrel-cage-rotor.

The rotor core is of laminated construction with slots suitably punched in for accommodating the rotor winding/rotor bars. The punched laminated are slacked and fitted directly on to a shaft in the case of small machines while in the case of large machines a stuck of annular punching of a suitable cross-sectional area are fitted on to a spider-web arrangement on the shaft.

1.2.1 Slip Ring Induction Motor

The winding of a wound-rotor is polyphase with coils placed in the slots of this rotor core. It is similar to that of the stator except that the number of slots is smaller and fewer turns per phase of a heavier conductor are used. The rotor is wound and connected in star with three leads brought out of the machine via slip-rings placed on the shaft. The slip-rings are tapped by means of copper-carbon brushes. Wound rotor construction is generally employed for large size machines to be used where the starting torque requirements are stringent, external resistance can be induced in the rotor circuit through slip-ring for reducing the starting current and simultaneously improving the starting torque.

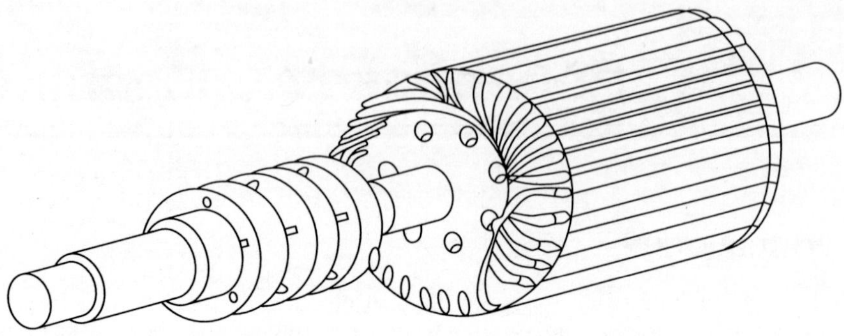

Fig. 1.1(*a*) Wound rotor for induction motor

1.2.2 Squirrel-Cage rotor

The squirrel-cage rotor has solid bars of conducting material placed in rotor slots and shorted through end rings on each side. In large machines alloyed copper bars are driven in the slots and are brazed onto copper end rings. Rotors up to 50 cm diameter usually have diecast aluminium bars wherein the end-rings are also cast simultaneously with the same material by using a suitable mould. This is an economical process and is generally employed in mass production of small size induction motors.

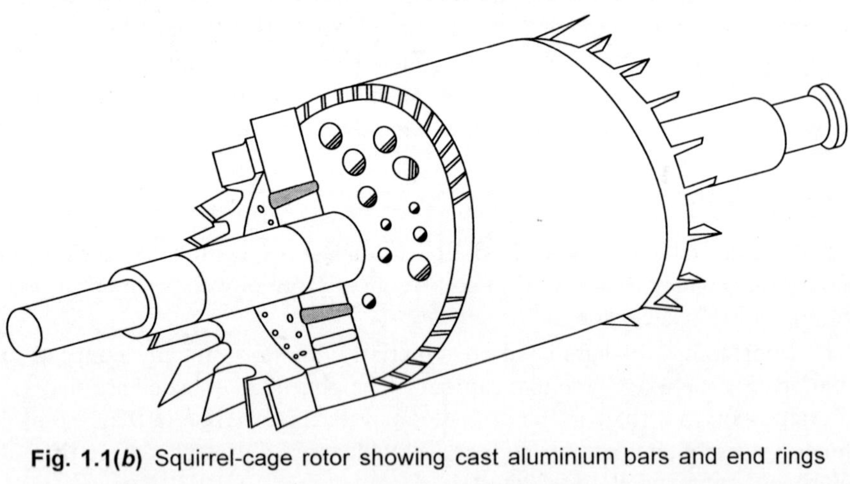

Fig. 1.1(*b*) Squirrel-cage rotor showing cast aluminium bars and end rings

The rotor circuit of a squirrel-cage machine cannot be tempered with and the machine have a low starting torque. While it has excellent running performance therefore it cannot be used where a high starting torque is required. The starting torque of a squirrel-cage motor can be improved by employing either a double-cage rotor or a deep-bar rotor. Such type of construction serves the purpose of a medium starting torque requirement. The rotor has a smaller number of slots than the stator and there must be a nonintegral multiple of stator slots so as to prevent magnetic locking of rotor and stator teeth of the time of starting. The slots in the induction machine are semi-enclosed so as to increase the permeance per pole so that magnetization current which is responsible for less than unity power of this motor is kept within limits.

The connection diagram of a 3-phase slip-ring induction motor with delta connected stator and star-connected rotor shown in below. The rotor winding is connected to slip rings which are shorted through external resistance at the time of starting. The resistance are cut-out as the motor attains full speed. The rotor of a squirrel-cage motor has permanently started bars.

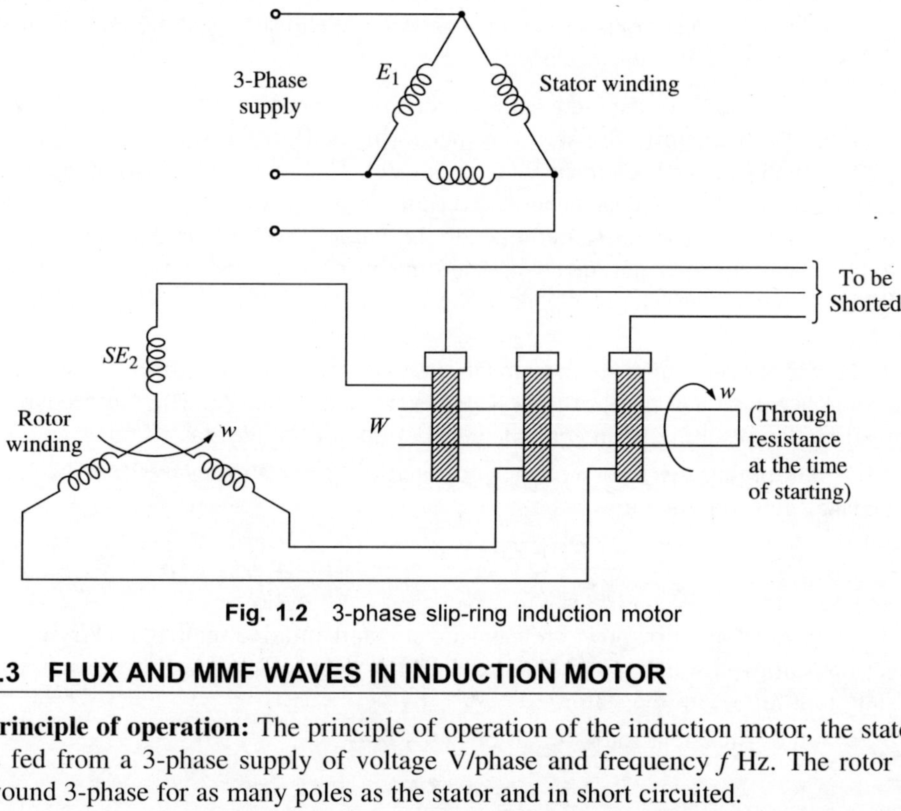

Fig. 1.2 3-phase slip-ring induction motor

1.3 FLUX AND MMF WAVES IN INDUCTION MOTOR

Principle of operation: The principle of operation of the induction motor, the stator is fed from a 3-phase supply of voltage V/phase and frequency f Hz. The rotor is wound 3-phase for as many poles as the stator and in short circuited.

$$V = E_1 = \pi\sqrt{2}\, k_{w1} N_{Ph1} f\, \phi_r \quad \text{...(1)}$$

where E_1 = stator induced emf/phase
k_{w1} = stator winding factor
N_{Ph1} = stator series turns/phase
ϕ_r = resultant air-gap flux/pole

1.4 DEVELOPMENT OF CIRCUIT MODEL (EQUIVALENT CIRCUIT)

The transformer analogy of induction motor. Certain facts established so far summarized below.

1. $\dfrac{E_1}{E_2} = a', \; \dfrac{I_2'}{I_2} = \dfrac{1}{a}$

where E_2 = standstill rotor emf.

Further I_2' flows in to the positive terminal of E_1 and I_2 flows out of the positive terminal of E_2. Also I_2 as seen from the stator is the current of stator frequency f and is in phase with I_2' the component of current drawn by the stator to balance the rotor mmf F_2.

2. Like in a transformer the magnetizing current component I_m of the stator current lags the stator induced emf E_1 by 90°.
3. The induction motor is not merely a transformer which changes voltages and current levels. It in fact behaves like a generalized transformer in which the frequency is also transformed in proportion to slip such that the rotor induced emf is SE_2 and rotor reactance is SX_2.

The circuit model of the inductance motor can now be drawn on a per phase basis as in Fig. 1.3(*a*) where in the series elements (lumped) of the stator resistance and leakage reactance have been included in the model. The transformer linking the stator and rotor circuits is an ideal generalized transformer in which the standstill rotor voltage E_2 and rotor current I_2 are linked to the stator quantities via transformer ratio a_1 while the frequency parameter appears in the rotor circuit through the slips a mechanical parameter. The mechanical power output appear at the shaft indicated in the figure.

The rotor circuit can be referred to the stator side by a two step process-modifying the rotor-circuit so that the turn-ratio becomes unity and then carrying out frequency transformation resulting in an equivalent rotor circuit of at the stator frequency.

By multiplying the rotor voltage by a and the rotor current by $1/a$. The rotor impedance gets modified to

$$\bar{Z}_2' = a^2R_2 + JSa^2X_2$$

or $$\bar{Z}_2' = R_2' + JSX_2';\ R_2' = a^2R_2,\ X_2' = a^2X_2 \qquad ...(i)$$

In this transformation power remains in variant the rotor circuit after carrying out this step is drawn in Fig. 1.3(*b*). This reduced the rotor to an equivalent rotor having a unity turn-ratio with the stator.

From the equivalent rotor circuit of Fig. 1.3(*b*)

$$I_2' = \frac{S\,\overline{E_1}}{R_2' + JS\,X_2'} \qquad ...(ii)$$

Dividing both numerator and denominator by S

$$I_2' = \frac{\overline{E_1}}{R_2'/S + J\,X_2'} \qquad ...(iii)$$

The simple trick refers the rotor circuit to the stator frequency. The modified rotor circuit in now drawn in Fig. 1.3(*c*) where in since both the rotor and station circuit have the same frequency. The ideal transformer is now a stationary unit ratio transformer. It is also noticed that in referring the rotor circuit to the stator frequency the reactances becomes constant (X_2') and the resistance becomes variable (R_2'/S). The transformation of Eq. (*iii*) is not power invariant (voltage changes while current remains the same) the power transferred to the secondary now account for both the rotor copper-law and mechanical power output (in electrical form) this is in contrast to Fig. 1.3(*a*) where mechanical power output is taken off via a shaft.

As a last step in equivalent circuit development, the ideal (ordinary) unit ratio transformer can now be dispensed with resulting in the circuit model of Fig. 1.3(*d*).

The representation of iron-loss in the stator can be heuristically introduced in the circuit model of Fig. 1.3(*d*) by placing a resistance R_1 in parallel with X_m as in the transformer circuit model. This circuit is drawn in Fig. 1.3(*e*).

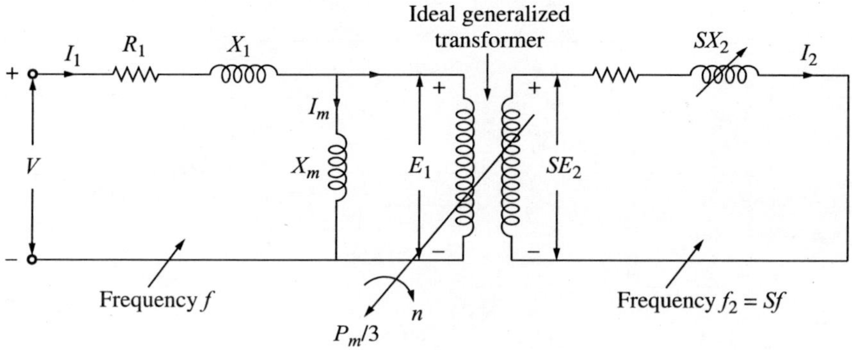

(*a*)

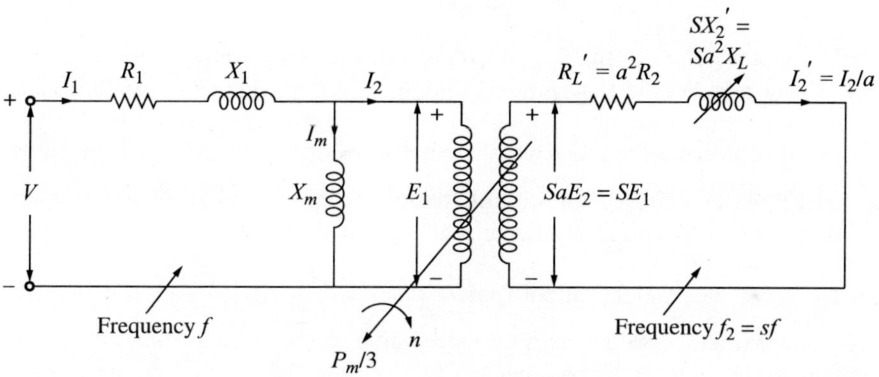

(*b*)

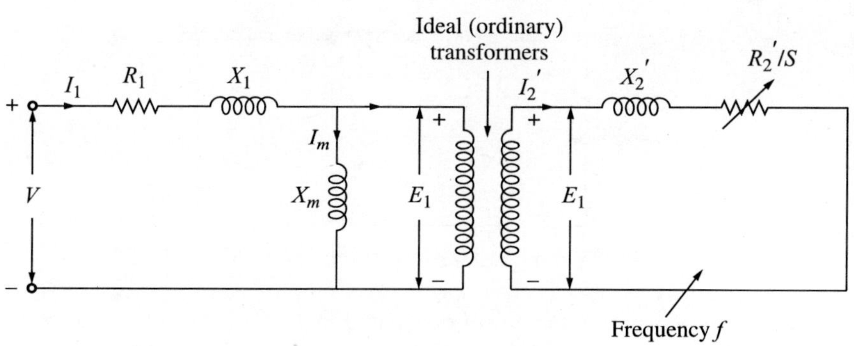

(*c*)

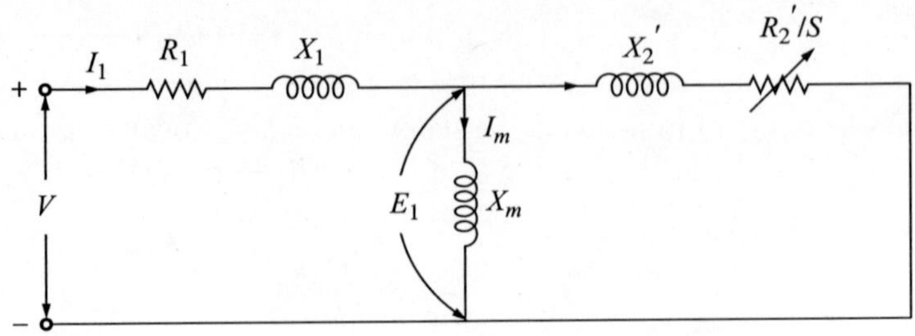

(d)

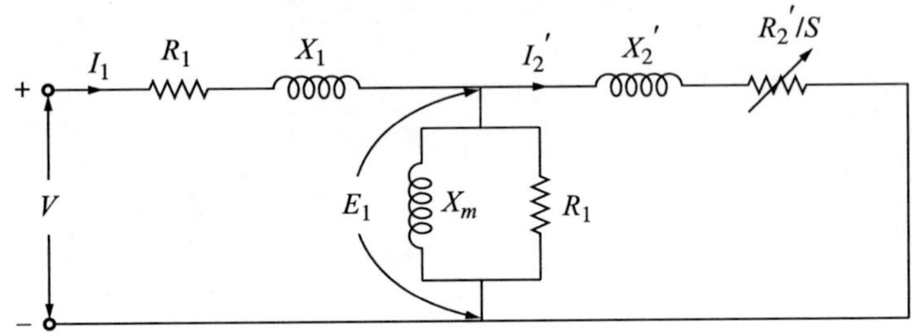

(e)

Fig. 1.3 Development of the circuit model (equivalent circuit) of induction motor

If R_2' is separated from R_2'/S to represent the rotor copper-loss as a separate entity. The circuit model can be drawn as in Fig. 1.3(*a*) in which the variable resistance $R_2'\left(\frac{1}{S}-1\right)$ represents the mechanical output in electrical form as will be shown in Fig. 1.3(*b*). After natively the circuit model of Fig. 1.3(*b*) (this corresponds) to Fig. 1.3(*d*). wherein the iron loss resistance R_i is omitted and this loss would be subtracted from the gross mechanical output (power absorbed by $R_2'\left(\frac{1}{S}-1\right)$ this amounts to certain approximation which is quite acceptable in the normal range of slip in an induction motor. Further the parameters of this circuit (which does not need the value of R_i) can be easily obtained by two non-loading tests to the described.

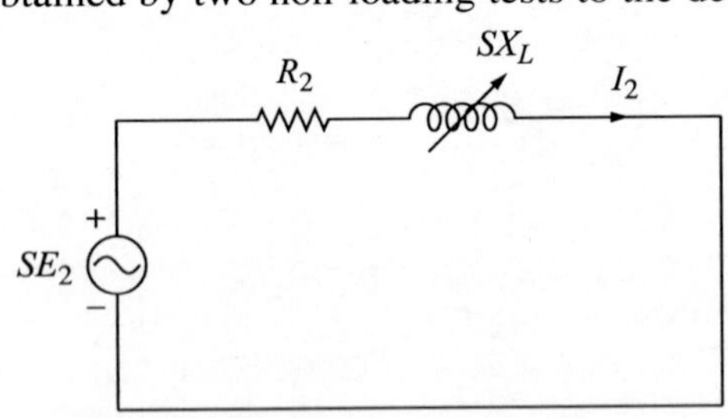

Fig. 1.4 Rotor circuit (frequency *Sf*)

It may be noted here that the power dissipated in $R_2'(1/S–1)$ includes the core loss which must be subtracted from it to obtain the gross mechanical power for getting net mechanical power output. The windage and friction loss must be further subtracted from it. The core loss and windage and friction loss together are lumped as rotational loss a both these losses occur when the motor is running. The rotational motor speed varies very little from no load to full load.

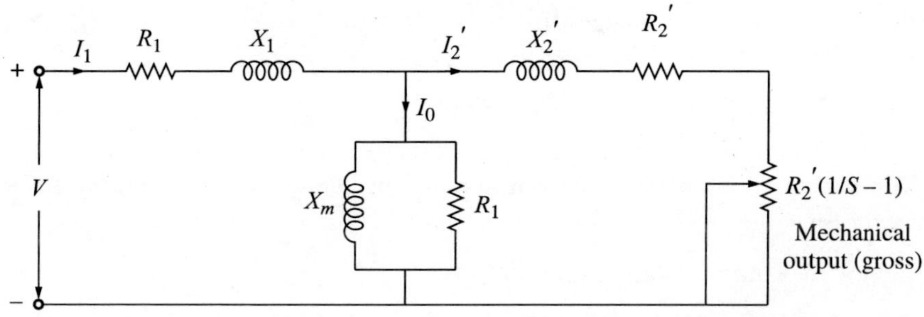

(a)

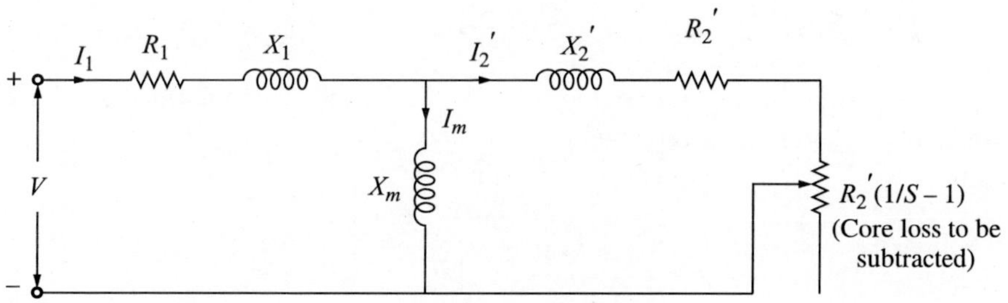

(b)

Fig. 1.5

Speed and Slip: An induction motor cannot run of synchronous speed. Let us consider that is rotor is rotating at relating synchronous speed under this condition there would be no cutting of flux by the rotor conductors, and there would be no generated voltage no current and no torque. The rotor speed in therefore slightly less than Ae synchronous speed. An induction motor may also be called as 'asynchronous motor' at it does not run at synchronous speed. The difference between the synchronous speed and the actual rotor speed is called the slip speed.

If N_s = synchronous speed in rpm

N_r = actual rotor speed in rpm

The slip speed = $N_s - N_r$ rpm ...(1)

It is denoted by S.

$$S = \frac{N_s - N_r}{N_s} \quad ...(2)$$

$$\text{Percentage slip} = \frac{N_s - N_r}{N_s} \times 100 \quad ...(3)$$

I other ward it

h_s = synchronous speed in rps

h_r = actual rotor speed in rps

Then $$S = \frac{h_s - h_r}{h_s} \qquad \text{...(4)}$$

$$\text{Percentage slip} = \frac{h_s - h_r}{h_s} \times 100 \qquad \text{...(5)}$$

Also $$S = \frac{w_s - w_r}{w_s} \qquad \text{...(6)}$$

The slip at full load varies from about 5 per cent for small motors to about 2 per cent for large motors.

Frequency of Rotor Voltage and Current

The frequency of current and voltage in the stator must be the same as the supply frequency given by

$$f = \frac{PN_s}{120} \qquad \text{...(1)}$$

The frequency in the rotor winding is variable and depends on the difference between the synchronous speed and the rotor speed. Hence the rotor frequency depends upon the slip. The rotor frequency is given by

$$f_r = \frac{P(N_s - N_r)}{120} \qquad \text{...(2)}$$

Division of Eq. (2) by Eq. (1) gives

$$\frac{f_r}{f} = \frac{N_s - N_r}{N_s} \qquad \text{...(3)}$$

But $$\frac{N_s - N_r}{N_s} = S$$

$\therefore$ $$f_r = sf$$

That is

rotor current frequency = per unit slip × supply frequency

Standstill condition $N_r = 0$

$$S = \frac{N_s - N_r}{N_s} = 1$$

and $$f_r = f$$

when we take $f = f_1$ and $f_r = f_2$

$$f_2 = sf_1$$

Rotor Current:

(*a*) Standstill conditions

Let E_2 = e.m.f. included per phase of the rotor at standstill

R_2 = Resistance per phase of the rotor

X_2 = reactance per phase of the rotor of standstill

$X_2 = 2\pi f_i L_2$

Z_2 = rotor impedance per phase of standstill

I_2 = rotor cannot per phase at standstill

$$Z_2 = R_2 + jX_2 \quad ...(1)$$

$$I_2 = \frac{E_2}{Z_2} \quad ...(2)$$

Power factor at standstill

$$\text{Cos } \phi_2 = \frac{R_2}{Z_2} = \frac{R_2}{\sqrt{R_2^2 + X_2^2}} \quad ...(3)$$

(*b*) Rotor current of slip *S*:

Induced emf per phase in the rotor winding at slip *S*

$$E_2 = SE_2 \quad ...(4)$$

Rotor winding resistance per phase = R_2

Rotor winding reactance per phase of slip *S* in

$$X_{2s} = 2\pi f_2 L = 2\pi(sf_1)\, L = sX_2 \quad ...(5)$$

Rotor winding impedance per phase at slip *s* is

$$Z_2 = R_2 + jX_{2s} = R_2 + jsX_2 \quad ...(6)$$

Rotor current at slip *s* is

$$I_{2s} = \frac{E_{2s}}{Z_{2s}} \quad ...(7)$$

Power factor at slip *s* is

$$\text{Cos } \phi\ 2s = \frac{R_2}{Z_{2s}} \quad ...(8)$$

Relation between Rotor Copper less and Rotor Input:

Z_d = developed torque = torque exerted on the rotor by rotating flux

n_s = synchronous speed (rps)

n_r = rotor speed (rps)

Power transferred from stator to rotor = air gap power P_g

$$P_g = w_s\, Z_d = 2\pi h_s\, \tau_d = \text{input power to rotor} \quad ...(1)$$

Total mechanical power developed by the rotor

$$P_{md} = w_r z_d = 2\pi\, h_r \tau_d\, w \quad ...(2)$$

Total I^2R loss in rotor = (power transferred from stator to rotor) – (total mechanical power developed by rotor)

$$P_{rc} = P_g - P_{md} = 2\pi(h_s - h_r)\, \tau_d \quad ...(3)$$

$$\therefore \quad \frac{\text{Total } I^2R \text{ loss in rotor}}{\text{Input power to rotor}} = \frac{2\pi(h_s - h_r)\tau_d}{2\pi h_s\, \tau_d} \quad ...(4)$$

$\therefore$ Rotor copper loss = sX rotor input

$$P_{rc} = sP_g = P_{ir} \quad ...(5)$$

The rotor copper loss is equal to slip times the rotor input (air-gap power). The term sP_g is known as slip because it is proportional to the slip for a given value of P_g. It is the portion of the air-gap power which is not converted into mechanical power

then rotor input = mechanical power developed + rotor copper loss

$$P_{ir} = P_{md} + P_{rc} \quad ...(6)$$

$$P_{rc} = S(P_{rd} + P_{rc})$$

$$P_{rc}(1 - s) = S\,P_{md}\,P_{rc} = \frac{S}{1-S} P_{rd} \quad ...(7)$$

rotor copper loss $= \dfrac{s}{1-s}$ × mechanical power developed by the rotor

$$P_g : P_{rc} : P_{md} = 1 : s : (1 - s) \quad ...(8)$$

There are three quantities may be bound from the slip and synchronous speed.

$$P_{rc} = sP_g \quad ...(9)$$

$$P_{md} = (1 - s)\,P_g \quad ...(10)$$

$$\tau_d = \frac{P_g}{w_s} \quad ...(11)$$

Developed torque τ_d: The developed torque or induced torque in a machine is defined as the torque generated by the internal electric-to-mechanical power common. The torque is also called the electromagnetic torque.

The developed torque is given by

$$\tau_d = \frac{\text{Mechanical power developed}}{\text{Mechanical angular velocity of the rotor}} = \frac{P_{md}}{w_r} \quad ...(1)$$

Since $P_{md} = (1 - s)\,P_g$

and $w_r = (1 - s)\,w_s$

$$\tau_d = \frac{(1-s)\,P_g}{(1-s)\,w_s} = \frac{P_g}{w_s} \quad ...(2)$$

Equation (2) indicate developed torque directly in terms of air-gap power P_g and synchronous speed w_s. Since w_s is constant and intendant of load conditions.

The developed torque is given by equation (2), the air gap power P_g is often called "the torque in synchronous watts".

Synchronous watt is the torque that develops power of 1 watt when the machine is running of synchronous speed.

Output power $P_0 = w_r\,\tau_{\text{load}}$

$$\tau_{\text{load}} = \frac{P_0}{w_r} = \frac{P_{md} - P_{rot}}{w_r}$$

Torque of an Induction Motor: Electrical power generated in rotor

$$= 3E_{2s}I_{2s}\cos\phi_{2s}\ \text{W}$$

$$= 3E_{2s}\cdot\frac{E_{2s}}{Z_{2s}}\cdot\frac{R_2}{Z_{2s}} = \frac{3E_{2s}^2R_2}{Z_{2s}^2}$$

$$= \frac{3s^2E_2^2R_2}{R_2^2+(sX_2)^2} \qquad ...(1)$$

All this power is dissipated as I^2R loss (copper loss) in the rotor circuit.

Input power to rotor $= 2\pi\, h_s\, \tau_d$

$$sX_2\, \pi h_s \tau_d = \frac{3S^2E_2^2R_2}{R_2^2+S^2X_2^2}$$

$$\tau_d = \frac{3E_2^2}{2\pi h_s}\cdot\frac{sR_2}{R_2^2+s^2X_2^2} \qquad ...(2)$$

$$\tau_d = \frac{ks\,E_2^2R_2}{R_2^2+s^2X_2^2} \qquad ...(3)$$

where $k = \dfrac{3}{2\pi h_s} = \dfrac{3}{w_s}$ = a constant ...(4)

Starting torque: At start, $s = 1$ therefore starting torque may be obtained by putting $s = 1$ in Eq. (2)

$$\tau_{\text{dst}} = \frac{3E_2^2R_2}{2\pi h_s(R_2^2+X_2^2)} \qquad ...(5)$$

The starting torque is also known as standstill torque.

Torque of synchronous speed: At synchronous speed, $s = 0$ and therefore $\tau_d = 0$, that is at synchronous speed, developed torque is zero

$$\frac{E_2}{E_1} = \frac{T_{e2}}{T_{e1}} \qquad ...(6)$$

$$E_2 = \frac{T_{e2}}{T_{e1}}\times E_1$$

is

$$\tau_d = \frac{3}{2\pi h_s}\left(\frac{T_{e2}}{T_{e1}}\right)^2 E_1^2\,\frac{sR_2}{R_2^2+s^2X_2^2} \qquad ...(7)$$

Let

$$\frac{3}{2\pi h_s}\left(\frac{T_{e2}}{T_{e1}}\right)^2 = k\ \text{(a constant)} \qquad ...(8)$$

$$\tau_d = \frac{k\,E_1^2sR_2}{R_2^2+s^2X_2^2} \qquad ...(9)$$

E_1 is nearly equal to V_1

$$\tau_d = \frac{kV_1^2 sR_2}{R_2^2 + s^2 X_2^2} \quad ...(10)$$

Starting torque is obtained by putting $s = 1$ in the above equation

$$\tau_{st} = \frac{kV_1^2 R_2}{R_2^2 + X_2^2} \quad ...(11)$$

$$\therefore \quad \tau_{st} \propto V_1^2 \quad ...(12)$$

That is the starting is proportional to the square of the stator applied voltage.

Condition for Maximum Torque: The value of torque when motor is running is given by

$$\tau_d = \frac{ksR_2 E_2^2}{R_2^2 S^2 X_2^2} \quad ...(1)$$

if the impedance of the stator winding is assumed to the negligible, the given supply voltage V_1, E_2 remains constant (1)

Let $\quad kE_2^2 = k_1$ (constant)

$$\tau_d = \frac{k_1 SR_2}{R_2^2 + S^2 X_2^2} \quad ...(2)$$

$$\tau_d = \frac{k_1 R_2}{\frac{R_2^2}{S} + SX_2^2} \quad ...(3)$$

$$= \frac{k_1 R_2}{\left(\frac{R_2}{\sqrt{S}}\right)^q + 2R_L X_2} \quad ...(4)$$

The developed torque τ_d will be maximum when the right-hand side of equation (4) is a maximum which is possible when

$$\frac{R_2}{\sqrt{S}} - X_2\sqrt{S} = 0$$

$$R_2 = SX_2 \quad ...(5)$$

$$R_2 = X_2 \quad ...(6)$$

The developed torque is a maximum when the rotor resistance per phase is equal to the rotor reactance per phase under running conditions.

The maximum torque is obtained by putting $SX_2 = R_2$ in the expression for torque Eq. (1)

$$\tau_{d\,max} = \frac{kSR_2 E_2^2}{R_2^2 + R_2^2} = \frac{kSE_2^2}{2R_2} = \frac{kSE_2^2}{2SX_2}$$

$$\tau_d = \frac{kE_2^2}{2X_2} \quad ...(7)$$

This relation shows that the maximum torque is independent of rotor resistance. If S_m = value of slip corresponding to maximum torque. Then from Eq. (5)

$$S_m = \frac{R_2}{X_2} \quad ...(8)$$

We have $N = N_s(1 - S)$

The speed of the rotor at maximum torque is

$$N_m = N_s(1 - S_m) \quad ...(9)$$

From Eq. (7) for maximum torque the following conditions can be drawn.

(*a*) Maximum torque is independent of rotor circuit resistance.

(*b*) Maximum torque varies inversely as standstill reactance of the rotor. Hence, for maximum torque, X_2 and therefore, the inductance of the rotor should be kept as small as possible.

(*c*) The slip at which the maximum torque depends upon the rotor resistance ($S_m = R_2/X_2$). Therefore by varying the resistance in the rotor circuit, maximum torque can be obtained at any desired slip or motor speed. If is to be noted that the resistance on the rotor circuit can only be varied in slipping rotors. In order to develop maximum torque at standstill, the rotor resistance must be high (equal to X_2) but to develop maximum torque under running condition the rotor resistance must be low.

Maximum Torque at Starting: To determine the condition for maximum torque at starting. Put $S = 1$ in Eq. (3) therefore, the starting torque will be a maximum when

$$\frac{R_2}{X_2} = S = 1$$

or

$$R_2 = X_2 \quad ...(10)$$

The rotor is not more than 1 or 2 percent of its leakage reactance for higher efficiency. In order to increase the starting torque extra resistance should be added to the rotor circuit at start to the rotor circuit of start and cut-off gradually as motor speeds up.

Full-Load Torque and Maximum Torque:

Let S = full-load slip of the motor

τ_{fl} = full-load torque

τ_{st} = starting torque

$$\tau_{fl} = \frac{kSR_2E_2^2}{R_2^2 + (SX_2)^2} \quad ...(1)$$

$$\tau_{max} = \frac{kE_2^2}{2X_2} \quad ...(2)$$

At starting, $S = 1$

$$\tau_{st} = \frac{kR_2E_2^2}{R_2^2 + X_2^2} \quad ...(3)$$

$$\therefore \qquad \frac{\tau_{fl}}{Z_{\max}} = \frac{kSR_2E_2^2}{R_2^2+(SX_2)^2} \bigg/ \frac{kE_2^2}{2X_2} = \frac{2SR_2X_2}{R_2^2+(SX_2)^2}$$

But $R_L = S_m X_2$ ...(4)

$$\therefore \qquad \frac{\tau_{fl}}{\tau_{\max}} = \frac{2S \cdot S_m X_2^2}{S_m^2X_2^2+S^2X_2^2}$$

$$\frac{\tau_{fl}}{\tau_{\max}} = \frac{2S\,S_m}{S^2+S_m^2}$$

Also

$$\frac{\tau_{st}}{\tau_{\max}} = \frac{k\,R_2E_2^2}{R_2^2+X_2^2} \bigg/ \frac{kE_2^2}{2X_2}$$

$$= \frac{2R_LX_L}{R_2^2+X_2^2} = \frac{2(S_mX_2)X_2}{(S_mX_2)^2+X_2^2}$$

$$\therefore \qquad \frac{\tau_{st}}{\tau_{\max}} = \frac{2S_m}{1+S_m^2} \qquad \text{...(6)}$$

Equation (6) can also be obtained from Eq. (5) by putting $S = 1$ in it

$$\frac{\tau_{st}}{\tau_{\max}} = \frac{2X\,LX_{sm}}{1+S_m^2} = \frac{2S_m}{1+S_m^2}$$

Also

$$\frac{\tau_{st}}{\tau_{fl}} = \frac{S^2+S_m^2}{S(1+S_m^2)} \qquad \text{...(7)}$$

1.5 TORQUE SLIP CHARACTERISTICS

The expression for torque-slip characteristic ($T_{(s)}$) is easily obtained by finding the Thevenin equivalent of the circuit to left of *ab* in Fig. 1.6(*a*).

$$\bar{Z}_{TH} = (R_1 + jX_1) \| jX_m = R_{TH} + j\,X_{TH}$$

$$V_{TH} = V\left[\frac{j\,X_m}{R_1 + j(X_1+X_m)}\right]$$

The circuit then reduces to Fig. 1.6(*b*) in which it is convenient to take V_{TH} as the reference voltage from Fig. 1.6(*b*).

$$I_2'^2 = \frac{V_{TH}^2}{(R_{TH}+R_2'/S)^2+(X_{TH}+X_2')^2} \qquad \text{...(1)}$$

$$T = \frac{3}{w_s} I_2'^2 (R_2'/S)$$

$$= \frac{3}{w_s}\frac{V_{TH}^2(R_2'/S)}{(R_{TH}+R_2'/S)^2+(X_{TH}+X_2')^2} \qquad \text{...(2)}$$

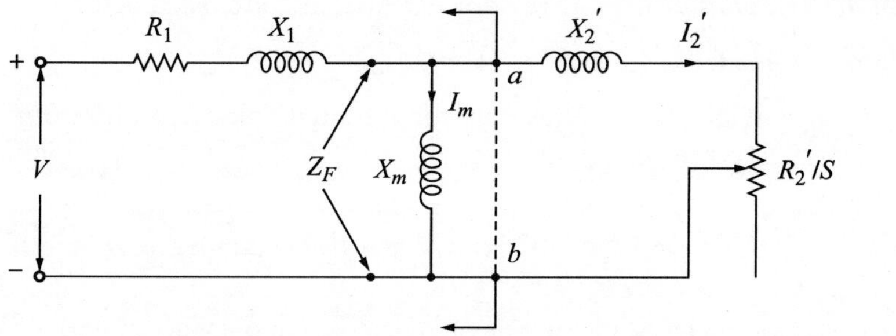

(a)

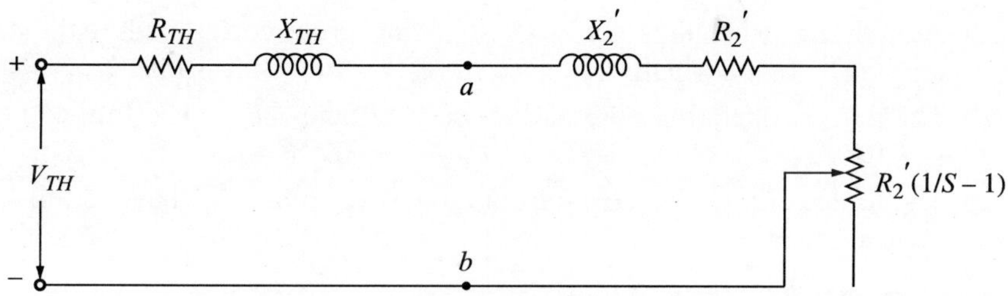

(b)

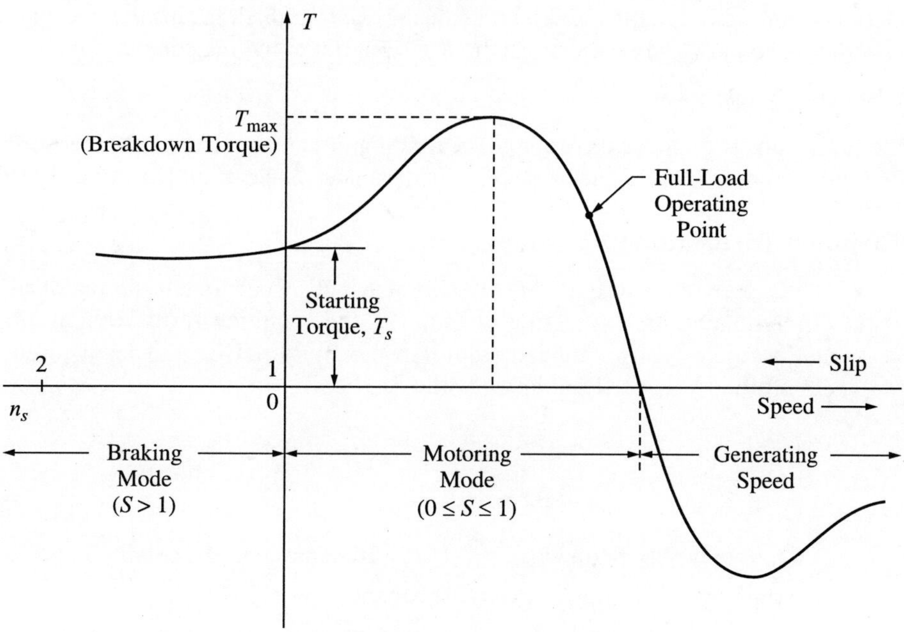

(c) Torque-slip characteristics

Fig. 1.6

Equation (2) is the expression for torque developed as a function of voltage and slip. For a given value of slip, torque is proportional to the square of voltage. The torque-slip characteristics at fixed (rated) voltage is plotted in Fig. 1.6(*c*).

Certain features of the torque-slip characteristics are listed below:

1. Motoring Mode: $0 \leq S \leq 1$

For this range of slip, the load resistance in the circuit model of Fig. 1.6(*b*) is positive, *i.e.*, mechanical power is output or torque developed is in the direction in which the rotor rotates. Also:

(*a*) Torque is zero at $s = 0$, as expected by qualitative reasoning advanced in previous section.

(*b*) The torque has a maximum value, called the breakdown torque (T_{BD}) at slip $S_{\max, T}$. The motor would decelerate to a holt if it is loaded with more than the breakdown torque.

(*c*) At $s = 1$, *i.e.*, when the rotor is stationary the torque corresponds to the starting torque, T_s. In a normally designed motor T_s is much less than T_{BD}.

(*d*) The normal operating point is located well below T_{BD}. The full-load slip is usually 2.8%.

(*e*) The torque-slip characteristic from no-load to some what beyond full-load is almost linear.

2. Generating Mode : $s < 0$

Negative slip implies rotor running at super-synchronous speed ($n > n_s$). The load resistance is negative in the circuit model of Fig. 1.6(*b*) which means that mechanical power must be put in while electrical power is put out at the machine terminals.

3. Braking Mode : $s > 1$

The motor runs in opposite direction to the rotating field (*i.e.*, n is negative), absorbing mechanical power (braking action) which is dissipated as heat in the rotor copper.

Maximum (Breakdown) Torque

- While maximum torque and the slip at which it occurs can be obtained by differentiating the expression of Eqn. (2), the condition for maximum torque can be more easily obtained from the maximum power transfer theorem of circuit theory as per Eqn. given below.

$$T = \frac{P_G}{w_s} = \frac{3 I_2'^2 (R_2'/S)}{w_s} \text{ Nm}$$

The torque is maximum when $I_2'^2 (R_2'/S)$ is maximum, *i.e.*, maximum power is absorbed by R_2'/S in Fig. 1.6(*b*). This condition is given as

$$\frac{R_2'}{S_{\max, T}} = \sqrt{R_{TH}^2 + (X_{TH} + X_2')^2}$$

(matching of impedance magnitudes)

$$S_{\max, T} = \frac{R_2'}{\sqrt{R_{TH}^2 + (X_{TH} + X_2')^2}} \quad ...(3)$$

Substituting in Eqn. (2) simplying

$$T_{max} = \frac{3}{w_s} \cdot \frac{0.5\, V_{TH}^2}{\left[R_{TH} + \sqrt{R_{TH}^2 + (X_{TH} + X_2')^2}\,\right]} \qquad ...(4)$$

- It is immediately observed that the maximum torque is independent of the rotor resistance (R_2') while the slip at which it occurs is directly proportional to it.

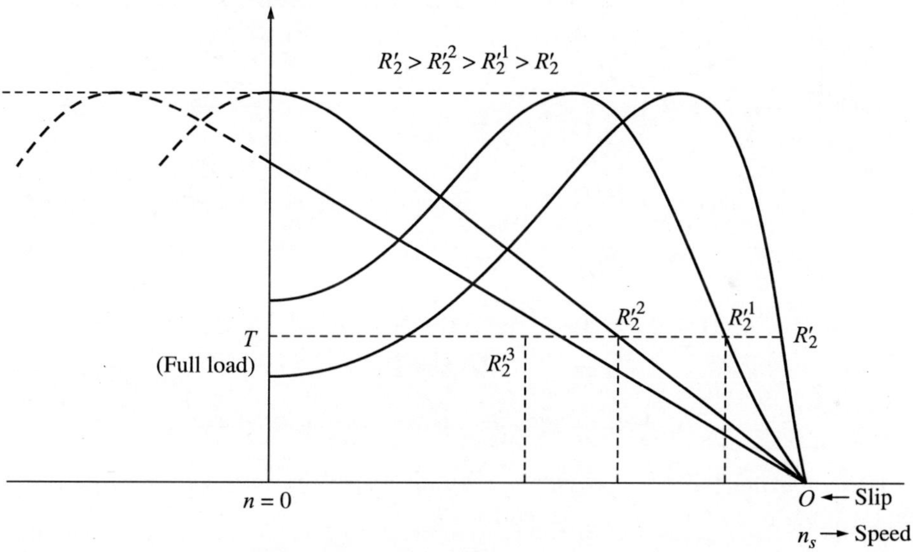

Fig. 1.6(*d*) Torque-slip characteristics of induction motor with increasing values of rotor resistance

- The torque-slip characteristic of a slip-ring induction motor can be easily modified by adding external resistance as shown in figure above by four such characteristics with a progressively increasing resistance in the rotor circuit.
- It may seen that as per Eqn. (4), the maximum torque remains unchanged while as per Eqn. (3) the slip at maximum torque proportionally increases as resistance is added to rotor circuit.
- Figure 1.6(*e*) below indicates the torque-slip profile at various supply voltages. Speed can also be controlled in this way by changing the stator voltage.
- It may be noted that the torque developed in an induction motor in proportional to the square of the terminal voltage.

Starting Torque

Let $S = 1$ in Eqn. (2)

$$T_{start} = \frac{3}{w_s} \cdot \frac{V_{TH}^2\, R_2'}{(R_{TH} + R_2')^2 + (X_{TH} + X_2')^2} \qquad ...(5)$$

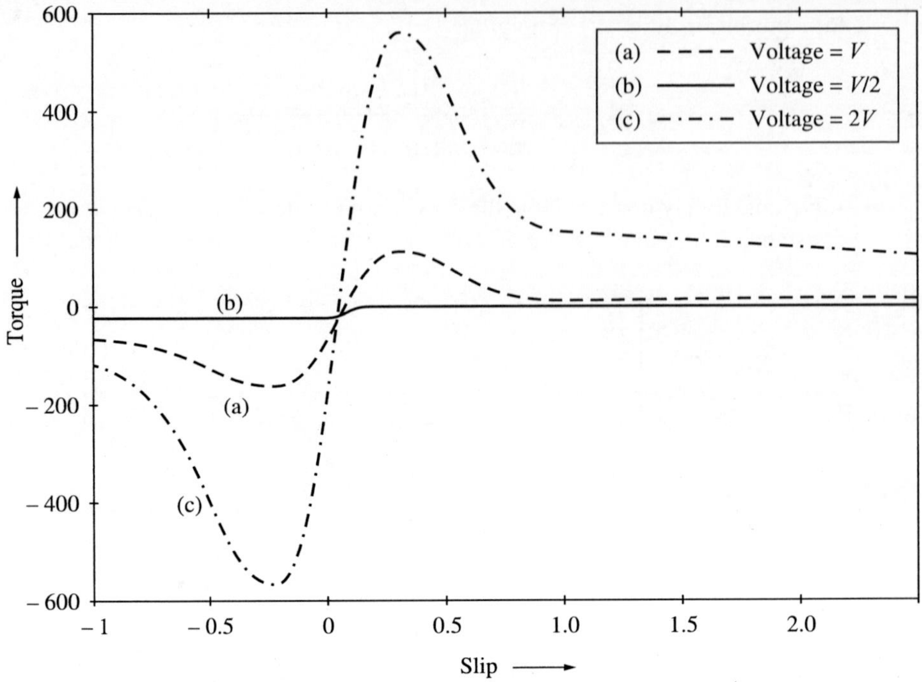

Fig. 1.6(*e*) Torque-slip curve at variable voltage

The starting torque increases by adding resistance in the rotor circuit. From Eqn. (3) the max. torque is achieved for ($S_{\max,\,T} = 1$)

$$R_2'\,(\text{total}) = \sqrt{R_{\text{TH}}^2 + (X_{\text{TH}} + X_2')^2} \quad \text{...(6)}$$

- At the same time the starting current will reduce. This indeed is the advantage of the slip-ring induction motor in which a high starting torque is obtained at low starting current.

An Approximation

Sometimes for getting a feel of the operational characteristics, it is convenient to assume the stator impedance to be negligible which leads to

$$R_{TH} = 0,\ X_{TH} = 0$$

in the Thevenin equivalent circuit of Fig. 1.6(*b*) so that $V_{\text{TH}} = V$. It then follows from Eqn. (1) to (6) that

$$I_2' = \frac{V}{\sqrt{(R_2'/S)^2 + X_2'^2}} \quad \text{...(7)}$$

$$T = \frac{3}{w_s} \frac{V^2 (R_2'/S)}{\sqrt{(R_2'/S)^2 + X_2'^2}} \quad \text{...(8)}$$

$$S_{\max,\,T} = \frac{R_2'}{X_2'} = \frac{R_2}{X_2} = \frac{\text{rotor resistance}}{\text{standstill reactance}} \qquad ...(9)$$

$$T_{\max} = \frac{3}{w_s} \cdot \left[\frac{0.5\,V^2}{X_2'}\right] \qquad ...(10)$$

and

$$T_{\text{start}} = \frac{3}{w_s} \cdot \frac{V^2 R_2'}{R_2'^2 + X_2'^2} \qquad ...(11)$$

- The maximum starting torque is achieved under the condition

$$R_2'(\text{total}) = X_2' \qquad ...(12)$$

and

$$T_{\text{start (max)}} = T_{\max} = \frac{3}{w_s}\left(\frac{0.5\,V^2}{X_2'}\right) \qquad ...(13)$$

Some Approximate Relationships at Low Slip

- A round the rated (full-load) speed, the slip of the induction is so small that

$$\frac{R_2'}{S} >> X_2'$$

so that X_2' can be altogether neglected in a simplified analysis. Equation (7) and (8) then simplify to

$$I_2' = \frac{SV}{R_2'} \qquad ...(14)$$

and

$$T = \frac{3}{w_s} \cdot S\frac{V^2}{R_2'} \qquad ...(15)$$

- It is immediately observed from Eq. (15) that the torque-slip relationship is nearly linear in the region of low slip which explains the linear shape of the characteristic as shown in Fig. 1.6(*c*).

Maximum Power Output

- Since the speed of the induction motor reduces with load, maximum mechanical power output does not correspond to the speed (slip) at which maximum torque is developed.
- For maximum mechanical power output, the condition is obtained from Fig. 1.6(*b*).

$$R_2'\left[\frac{1}{S} - 1\right] = \sqrt{(R_{TH} + R_2')^2 + (X_{TH} + X_2')^2} \qquad ...(16)$$

- The maximum power output can then be found corresponding to the slip defined by Eqn. (16). However this condition corresponds to very low efficiency and very large current and is well beyond the normal operating region of the motor.

1.6 DEEP-BAR/DOUBLE-CAGE ROTOR

The chief advantage of the slip-ring induction motors compared to the squirrel-cage one lies in the fact that while its rotor designed with low resistance to give good running performance (high efficiency). Excellent starting characteristic (low starting current, low slip high starting torque, etc.) is simply achieved by adding an external resistance in the rotor circuit at the time of starting which is then output gradually while the rotor reaches normal speed. The rotor circuit in the squirrel-cage motor, however, cannot be tempered with so that while its resistance in designed to give excellent running performance. It has high starting current and low starting torque which is further impaired by a reduced-voltage starting employed to limit the starting current (starting torque is proportional to the square of the voltage applied to the motor stator). The attractive qualities of low-cost, ruggedness and maintenance free operation of the squirrel-cage motor has impelled designers to find ways of improving its starting characteristic without sacrificing heavily its excellent running performance.

The fact that the rotor currents are of stator frequency (50 Hz) at the time of starting while this frequency reduces to $f_2 = Sf$ (may be as low as 2.5 Hz) under running condition can be exploited as it causes automatic variation of rotor resistance from a high value of starting (50 Hz resistance) to a low value under running (about 2.5 Hz resistance). This phenomenon is basically the skin and proximity effect which occurs an any conductor carrying alternating current. For the conductor cross-sectional shape (round or square) normally employed for rotor bars, this variation is not prominent enough to give low starting current and high starting torque. To enhance the variation in the effective resistance of rotor bars, deep bar. Conductors or double-cage rotor are employed.

1.6.1 Deep-Bar Rotor

In this type of construction bars of narrow width are laid down in deep semi, enclosed slots shown in below, the magnetic leakage flux pattern set up by the bar current is indicated in dotted lines in the figure. The rotor bar can be imagined to be compared of elementary strips in parallel-topmost and bottom most strips are shown in below. It is easily seen that a much larger flux links the bottom elementary strip compared to the top elementary strip as a consequence the starting reactance (50 Hz reactance) for the bottom strip is much larger than that of the top strip, it then follows that the current in the top strip is much more than the current in the bottom strip and further the top-strip current some what leads the bottom-strip current because of its lower reactance. Similar arguments when applied to other elementary strips would reveal that the current is unevenly distributed over the bar cross-section with the current demits progressively increasing while moving upwards from the bottom strip. Non-uniform current distribution comes greater ohmic loss meaning thereby that the effective bar resistance becomes much more than its dc resistance as the rotor speeds upto a value core to synchronous, the frequency of rotor current ($f_2 = Sf$) becomes very low. The reactances of various elementary strips of this low frequency become almost equal and the current

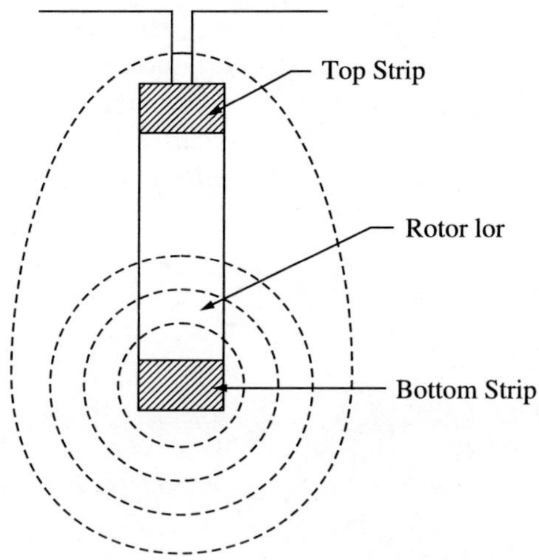

Fig. 1.7 Deep-bar rotor and leakage flux pattern

density over the conductor cross-section becomes nearly uniform. So that it offers a resistance almost equal to its dc value by choice of bar cross-sectional dimension, it is possible to obtain a starting rotor resistance (50 Hz resistance) to be many times the running rotor resistance (almost dc value). A deep-bar rotor therefore, has a low starting current and a high starting torque and a low running resistance which means that it can satisfactorily meet both the desired starting and running performances. Since the net rotor reactance at standstill is somewhat higher than in a normal bar design, the breakdown torque is somewhat lower. The torque-slip characteristic of a deep-bar rotor compared to a normal bar design with low and high rotor resistance is shown in Fig. 1.8.

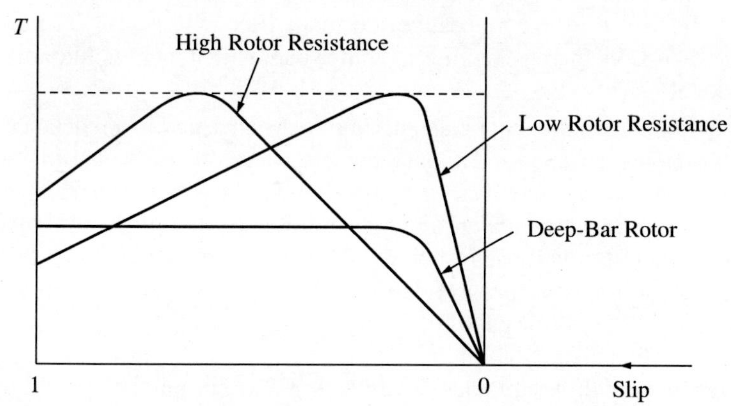

Fig. 1.8 Torque-slip of deep bar rotor

1.6.2 Double-Cage Rotor

A rotor design, which though more expensive gives still better starting and running performances than the deep-bar design is the double-cage rotor. The squirrel-cage winding in this design cavity of two layer of bar short-circuited by end rings. The upper bars have a smaller cross-sectional area than the lower bars and consequently a higher resistance. The slots accommodating the two sets of bars are joined by a constriction shown in Fig. 1.9.

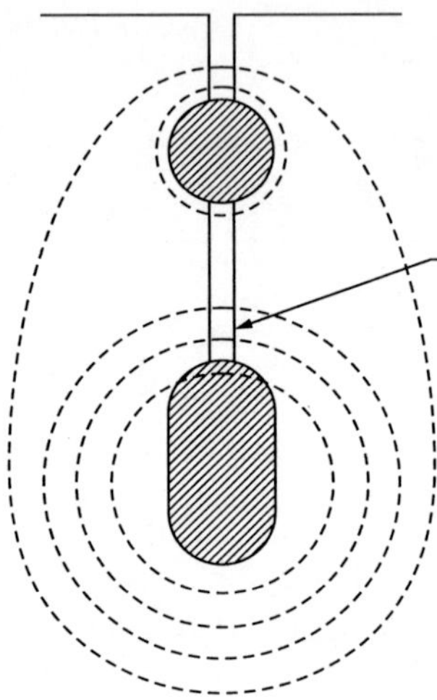

Fig. 1.9 Slot-leakage flux pattern for double-cage rotor

Which also shown the slot-leakage flux pattern for, the double-cage rotor by argument, similar to than prevented for these deep bar rotor construction. It is seen that the upper bars have a much lower leakage flux linkage and therefore a much lower reactance. Furthermore, the self leakage flux linking the upper/lower bars can be controlled by he dimension of the air construction. The constriction is also necessary because of the fact that in its absence the main flux will return via the iron path between the two slots thereby mining the outer bars which then would not contribute to torque development.

It is then found that the outer cage has high resistance and low reactance while the inner-cage has low resistance and high reactance. Therefore, in the starting, the current is mainly confined to the outer cage with a consequent decrease in starting current and an increase in starting torque under running condition the reactance difference between the two cages evens out because of low frequency of rotor currents such that there act to conduct current in inverse proportion to their dc resistance and as a group present a low resistance rotor to the air gap flux giving an excellent running performance. Since the starting current is mainly combined to the upper cage, this design is susceptible to frequent starting which would cause overheating and burning out of the upper high resistance cage.

Another type of a *double-cage rotor* construction in which the two cages are placed in staggered slots shown in Fig. 1.10.

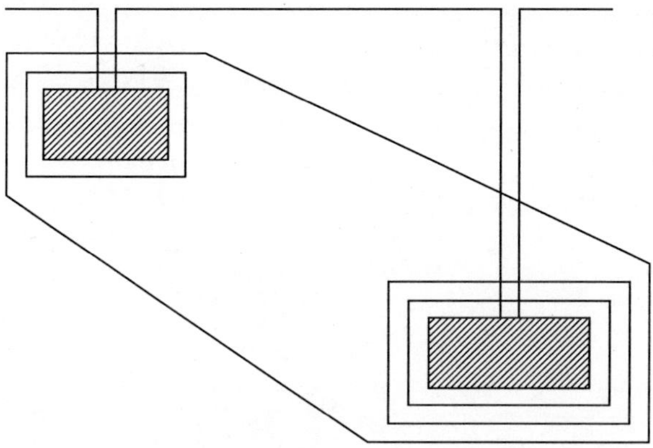

Double-cage rotor

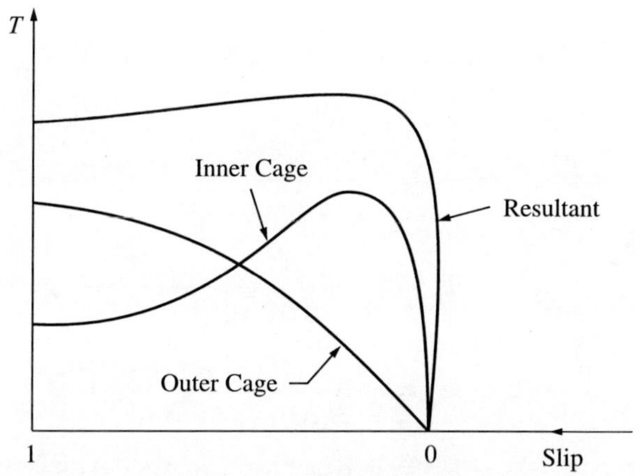

Torque-slip characteristic-double-cage induction rotor

Fig. 1.10

The two cages though somewhat coupled magnetically can be treated as Independent for Simplified yet fairly accurate analysis. The approximate circuit model of the double-cage rotor on the assumption in given below. The resultant torque-slip Characteristic as obtained by Summation of the torque developed by the inner and outer cage show in Fig. 1.10.

It will be evident to the reader by now that for low starting torque requirement which is the case with a vast majority of induction motor application the low-cost ordinary squirrel-cage construction in employed. A deep bar rotor construction is adopted for higher starting torque application and double cage construction for still higher starting torque needs.

Example 1.1

The impedances at standstill of the inner and outer cages of a double cage rotor are (0.01 + J 0.5) Ω and (0.05 + J 0.1) Ω respectively. The stator impedance may be assumed to be negligible.

Calculate the ratio of the torque due to the two cages.

(i) at starting (ii) when running with a slip of 5%.

Solution. Formula: $T_s = \frac{3}{w_s} \cdot \frac{V'^2 R_2^2}{R_2^2 + X_2^2}$ (rotor side)

where V' = rotor induced emf.

Substituting values

$$T_{so} = \frac{3}{w_s} \times \frac{V'^2 (0.05)}{(0.05)^2 + (0.1)^2}$$

$$T_{si} = \frac{3}{w_s} \times \frac{V_1'^2 (0.01)}{(0.01)^2 + (0.5)^2}$$

$$\frac{T_{so}}{T_{si}} = \frac{(0.01)^2 + (0.5)^2}{(0.05)^2 + (0.1)^2} \times \left(\frac{0.05}{0.01}\right) = 100$$

Formula:

$$T = \frac{3}{w_s} \cdot \frac{V_1'^2 (R_2/S)}{(R_2/S)^2 + X_2^2}$$

$$T_o = \frac{3}{w_s} \cdot \frac{V'^2 (0.05/0.05)}{(0.05/0.05)^2 + (0.1)^2}$$

$$T_i = \frac{3}{w_s} \cdot \frac{[V'^2 (0.01/0.05)]}{(0.01/0.05)^2 + (0.3)^2}$$

$$\frac{T_o}{T_i} = \frac{(0.01/0.05)^2 + (0.5)^2}{(0.05/0.05)^2 + (0.1)^2} \times \frac{0.05}{0.01}$$

$$= 1.436$$

The outer cage contributes 100 times more torque than the inner cage at starting while it contributes only 1.435 times during running.

A cage motor with two (or more) separate stator windings of different pole numbers can run of two (or more) integrally related speeds. Certain change pole motors have a single winding, the connections of which can be changed to give different pole numbers, but separate windings appropriately matched to the load. Conditions, may be preferred for such loads as lifts and hoists where a speed change must be made without losing control of the motor separate windings are however difficult to accommodate.

The coil pitch of a stator winding is naturally fixed, but its electrical span depends on the number of poles. A coil pitch one-eight of the circumference provides full-pitch coils for an 8-pole connection. Two-thirds for a 6-pole and one-half for a 4-pole connection too narrow a span is to be avoided and for a pole-change winding for a 2/1 speed ratio a possible arrangement is a coil span of $1\frac{1}{3}$ pole-pitch for the larger and 2/3 for the smaller pole number. In each case with a coil-span factor of 0.87, the spans may in practice be nearer to 1 and ½ (coil-span factors 1.0 and 0.71). In order to avoid excessive leakage reactance in the lower-speed connection.

Operating Conditions

P_1 and P_2 indicate the smaller and larger pole numbers. The ratio of gap flux densities B and that of the pull-out torque M are

$$\frac{B_1}{B_2} = \frac{N_2\,kw_2\,P_1}{N_1\,kw_1\,P_2} \cdot \frac{V_1}{V_2}$$

and
$$\frac{M_1}{M_2} = \frac{(B_1\,kw_1)^2}{(B_2\,kw_2)^2} \cdot \frac{P_2}{P_1}$$

where N, kw and V refer to the turns in series per phase, the winding factor and the phase voltage. The relative maximum output powers are $P_1/P_2 = (M_1/M_2)\,(P_2/P_1)$. Neglecting saturation, the magnetizing and short circuit current ratios are

$$\frac{I_{M1}}{I_{M2}} = \frac{B_1 P_1}{B_2 P_2} \cdot \frac{N_2 kw_2}{N_1 kw_1}$$

and
$$\frac{I_{sc1}}{I_{sc2}} = \frac{N_2^2}{N_1^2} \cdot \frac{V_1}{V_2}$$

for the letter ignoring change in overhang leakage if the same phase voltage is applied for each connection, then $V_2 = V_1$ the ratios are only approximate the magnetising-current-ratio being 10-20% greater and the short-circuit current ratio 10-15% smaller than as given because of the changed mmf pattern and the different conditions of saturation. The choice of winding arrangement depends on the load two cases are considered. Others (e.g. for far ventilation drives) may sometimes be more opposite.

Equivalent Circuit of a Double-cage Induction Motor

Let R_1 = resistance per phase of stator
X_1 = reactance per phase of stator
$R_2'o$ = resistance per phase of outer cage referred to stator
$X_2'o$ = Standstill leakage reactance per phase of outer cage referred to stator
R_{2i} = resistance per phase of the inner cage referred to stator
X_{2i} = standstill leakage reactance per phase of the inner cage referred to stator
S = fractional slip

If the main flux completely links both the cages, the impedance of the two cages can be considered in parallel.

The equivalent circuit of the double-cage induction motor of slip S is shown in above figure. If the shunt branches containing R_0 and X_0 are neglected. The equivalent circuit is simplified to that shown in Fig. 1.12.

Approximate equivalent circuit of a double-cage induction motor with magnetising current neglected.

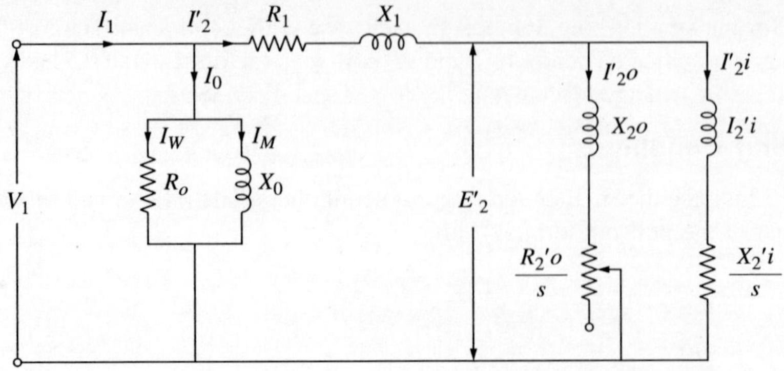

Fig. 1.11 Equivalent Circuit of a Double-cage Inductor Motor

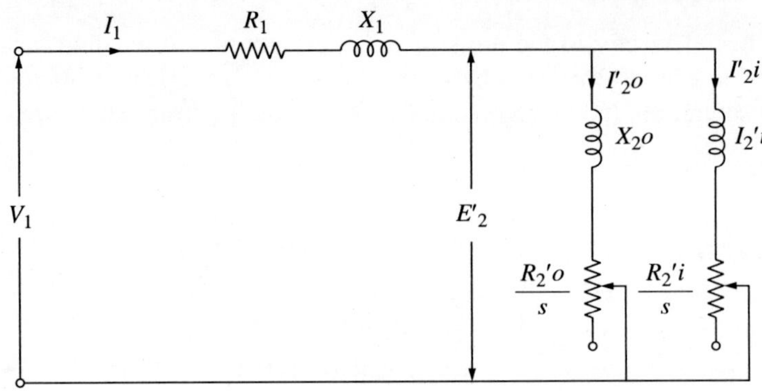

Fig. 1.12

At slip S, the outer-cage impedance, $Z'_{2o} = \dfrac{R'_{2o}}{j} + jX'_{2o}$

At slip S, the inner-cage impedance, $Z_{2i} = \dfrac{R_{2i}}{j} + jX'_{2i}$

The impedance of the stator, $Z_1 = R_1 + jX_1$

Equivalent impedance per phase of the motor referred to stator

$$Z_{e1} = Z_1 + (Z'_{2o} \| Z'_{2i})$$

$$\therefore \qquad Z_{e1} = R_1 + jX_1 + \frac{1}{\dfrac{1}{Z'_{2o}} + \dfrac{1}{Z'_{2i}}}$$

$$= X_1 + \frac{Z'_{2o} Z'_{2i}}{Z'_{2o} + Z'_{2i}} \qquad \text{...(1)}$$

Current through the outer cage

$$I_2' = \frac{E_2}{Z'_{2o}} \qquad \text{...(2)}$$

Current through the inner cage

$$I'_{2i} = \frac{E_2}{Z_{2i}} \qquad \text{...(3)}$$

The rotor current (referred to the stator) is equal to the phasor sum of the current through the outer and inner cages.

$$I_2' = I_{2o}' + I_{2i} \qquad ...(4)$$

Torque-slip Characteristics of a Double-cage Induction Motor:

The two cages develop two separate torques. The total torque of the motor is equal to the sum of the two cage torques. The torque slip characteristics of the two cages are shown in Fig. 1.13.

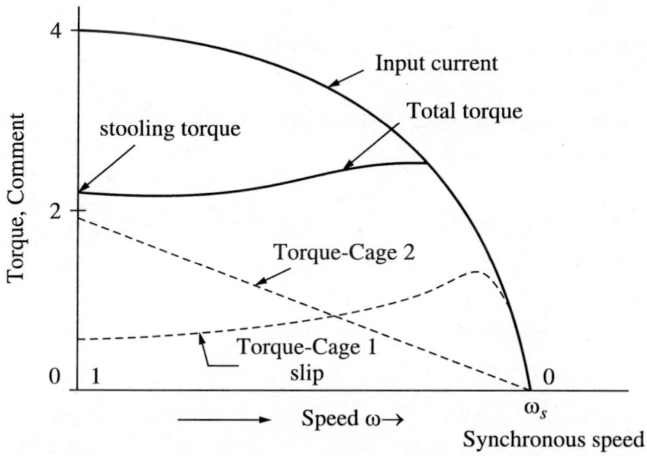

Fig. 1.13 Torque and current characteristic of a double-cage induction motor

The resultant torque-speed characteristics can be modified according to the requirement. This is done by modifying the individual cage resistances and leakage reactances. The resistance can be changed by changing the area of cross-section of bars. The leakage reactance can be changed by changing the width of the slot opening and the depth of the inner cage.

Comparison of Cage Torques

Power developed per phase by the outer cage

$$P_{do} = I_2^{2\prime} \frac{R_{2o}'}{S} \qquad ...(1)$$

Power developed per phase by the inner cage

$$P_{di} = I_{2i}^2 \frac{R_{2i}'}{S} \qquad ...(2)$$

Power developed per phase by both the cages

$$P_d = P_{do} + P_{di} = (I_2')^2 \frac{R_{2o}'}{S} + (I_{2i}')^2 \frac{R_{2i}'}{S} \qquad ...(3)$$

From the equivalent circuit of the double-cage motor

$$I_{2o}' = \frac{E_2'}{Z_{2o}'} \qquad ...(4)$$

$$I'_{2i} = \frac{E'_2}{Z'_{2i}} \quad ...(5)$$

$$Z'_{2o} = \sqrt{\left(\frac{R'_{2o}}{S}\right)^2 + (X'_{2o})^2} \quad ...(6)$$

$$Z'_{2i} = \sqrt{\left(\frac{R'_{2i}}{S}\right)^2 + (X'_{2i})^2} \quad ...(7)$$

If τ_{do} = torque developed by the router cage

τ_{di} = torque developed by the inner cage

τ_d = total torque developed by the two cages

$P_d = (2\pi n_s)\tau_d$

$$\tau_d = \frac{P_d}{2\pi n_s} = \frac{1}{2\pi n_s}\left[(I'_{2o})^2 \frac{R'_{2o}}{S} + (I'_{2i})^2 \frac{R'_{2i}}{S}\right] \quad ...(8)$$

$$\frac{\tau_{do}}{\tau_{di}} = \frac{\left(\frac{R'_{2o}}{S}\right)^2 + (X'_{2o})^2}{\left(\frac{R_{2i}}{S}\right)^2 + (X'_{2i})^2} \quad ...(9)$$

Effect of space harmonics on three-phase induction motor performance: The air-gap flux set up by the three-phase stator windings. Carrying sinusoidal. Current is of nonsinusoidal wave shape due to fourier series analysis, any nonsinusoidal flux is equivalent to be combination of a number of sinusoidal fluxes of fundamental and higher order harmonics. All even harmonics (2, 4, 6 ...) are about in fourier series. A nonsinusoidal flux can be resolved into fluxes of fundamental and higher-order odd harmonics (3rd, 5th, 7th, 11th, 13th, etc.)

The third harmonic flux waves produced by each of the three-phase neutralize one another. Therefore the resultant air gap flux is free from triples (that is, third and its multiplies 3, 9, etc.) harmonics. This is due to the fact that the third harmonics in the flux wave of all the three phases are in space phase. But differ in time phase by 120°, space harmonics fluxes are produced by windings, slotting, magnetic saturation, inequalities in the air gap length, etc. There harmonic fluxes induced voltage and circulate harmonic currents in the rotor windings. There harmonic current in the rotor interact with the harmonic fluxes to produce harmonic torques, vibrations and noise.

Harmonic induction torques: A 3-phase winding carrying sinusoidal currents produces space harmonics of the order

$$h = 6k \pm 1$$

where k is a positive integer (1, 2, 3 ...) the synchronous speed of the hth harmonic in $(1/h)$ times the speed of the fundamental wave the space harmonic waves rotate in the same direction as the fundamental wave. It $h = 6k + 1$, and in the opposite direction if $h = 6k - 1$.

A space harmonic wave of order h is equivalent to a machine with number of poles equal to ($h \times$ number of poles of the stator).

Therefore, the synchronous speed of the nth space harmonic wave is

$$h_s(h) = \frac{h_s}{h} = \frac{120f}{h \times P}$$

where f = supply frequency

P = number of poles of the stator

This, for k = 1, a 3-phase winding will produce predominant backward rotating fifth harmonic rotating at a speed of (1/5) of synchronous speed and forward rotating seventh harmonic rotating of a speed of (1/7) of synchronous speed. These harmonics above will have little effect on the operation of the motor. The torque-speed characteristics for the fundamental flux and fifth and seventh space harmonic flux are shown in figure below. The fifth and seventh harmonic torques have the same general shape as that of the fundamental.

Since fifth harmonic flux rotating in the directions opposite to the rotation of the rotor, the fifth harmonic torque opposites the fundamental component torque. In otherwords, the fifth harmonic flux produces a braking torques the seventh harmonic flux rotates in the same direction as the fundamental flux. Therefore, the seventh harmonic induction torque aids the fundamental component torque. The resultant torque-speed characteristic will be the combination of the fundamental fifth and seventh harmonic characteristics as shown in below. The resultant torque-speed characteristic has two dips, one near (1/5) of synchronous speed occur in the negative direction of the motor rotation. The dip near (1/7) of synchronous speed in more important, figure below also shows load-torque speed characteristic. If the motor torque is developed due to the fundamental flux alone. The motor will accelerate to the point L which is the interaction of the load torque characteristic and the motor torque-speed curve. Due to the presence of seventh harmonic flux torque, the load torque curve intersects the motor torque speed characteristic at point A. The seventh harmonic flux torque curve has a negative slope at point A stable running condition over. The torque range between the maximum and minimum points results. The motor torque falls below the load torque. At this stage the motor will not accelerate up to its normal speed, but will remain running at a speed which is nearly (1/7) of its normal speed and the operating point would be A. This tendency of the motor to run of a stable speed as low as one-seventh of the normal speed N_s and being unable to pick up its normal speed is known as crawling of the motor.

Crawling can be reduced by reducing fifth and seventh harmonics. This can be done by using a chorded (or short pitched) winding.

Cogging or Magnetic Locking: When full voltage is applied to the stator winding, the rotor of a 3-phase cage induction motor fails to start. This happens when the number of stator and rotor slots are either equal or have an integral ratio with the number of stator slots equal to or an integral multiple of rotor slots, strong alignment forces are produced between stator and rotor at the instant of starting there forces may create an alignment torque greater than the accelerating torque with consequent failure of the motor to start. This phenomenon of magnetic locking between stator and rotor teeth is called cogging on teeth locking.

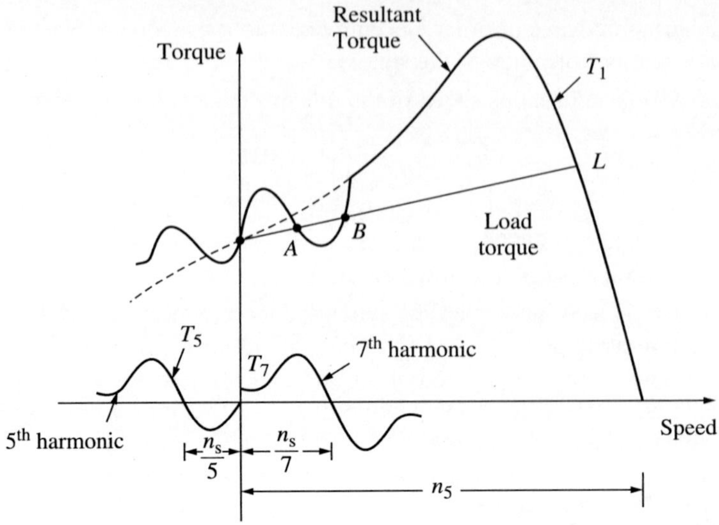

Fig. 1.14 Torque-speed characteristics of 3-phase induction motor showing the effect of space harmonic asynchronous torque

The reluctance of the magnetic path is minimum when the stator and rotor teeth face each other. Under this condition then is a magnetic locking between stator load rotor teeth.

To reduce or eliminate cogging the number of stator slots are never made equal to have an integral ratio. Cogging can also be reduced by using skewed rotor.

Cogging and crawling are much less prominent in wound rotor motors because of their higher starting torques.

1.7 EMF INJECTION IN ROTOR CIRCUIT OF SLIP RING INDUCTION MOTOR

The basic concepts regarding the voltage injection into the rotor circuit of a wound-rotor induction motor for its speed (and power factor) control are presented.

It is given in Eq. (17) that power transferred from stator to rotor is

$$P_g = sP_g + (1 - S)\, P_g \qquad \text{...(17)}$$

Here $(1 - S)\, P_g$ represent the mechanical power developed in the rotor and slip power sP_g appear as rotor ohmic loss $(I_2^2 r_2)$ if rotor is short-circuited. However, instead of wasting the slip frequency power sP_g in the rotor circuit, it can be converted by various schemes to useful power which is either returned to the supply or added to the motor shaft.

The most useful method of recovering the slip power sP_g and controlling the speed thereby, involves the injection of an emf into the rotor circuit from some suitable external sources. It should be understood that the injected emf must be at slip frequency at all the rotor speed. Note that when the emf injected in the rotor circuit, the wound-rotor induction motor become a doubly-excited induction machine.

The effect of an injected emf into the rotor circuit of induction motor, can be better understood with the help of its phasor diagrams.

Figure 1.15(*a*) show the phasor diagram of an induction motor running at slip s and with its rotor-short circuited. In order to concentrate more on the effects of emf injection. The primary impedance drop has been neglected. If the load torque on the induction motor is assumed constant and no load rotational losses are neglected, then per phase electromagnetic torque T_e is

$$T_e = \frac{E_2}{w_s} I_2 \cos\theta_2 = \left[\frac{E_2}{2\Omega n_s}\right] I_2 \cos\theta_2 = \text{Load torque}$$

Since, E_2 and n_s are constant, $I_2 \cos\theta_2$ must be treated as constant for a constant-torque load. Here E_2 is the emf induced in the rotor phase winding by the main air gap flux at standstill and sE_2 is its magnitude at any slip 5. Synchronous speed n_s is in *rps*. Rotor current I_2 lags sE_2 due to rotor leakage reactance and therefore, R_1 lags v_1 giving a lagging power factor operation. In Fig. 1.15(*b*), an emf E_j is injected in phase opposition to standstill rotor emf E_2. Since the speed and therefore the slip can't change suddenly due to rotor inertia, the immediate effect of voltage injection is to decrease the rotor current from

$$\frac{sE_2}{Z_2} \text{ to } \frac{sE_2 - E_j}{Z_2}$$

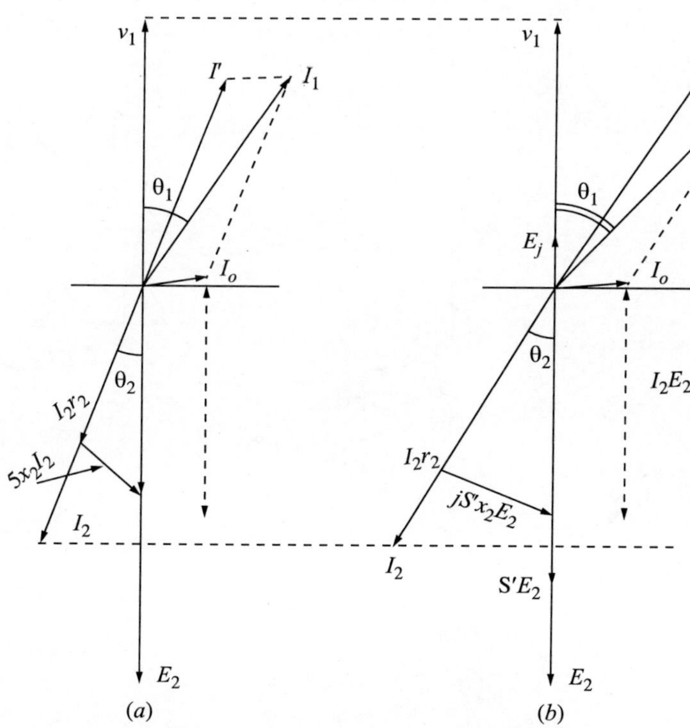

(*a*)
Ordinary IM phasor diagram

(*b*)
(*i*) Speed = h_r, slip = s

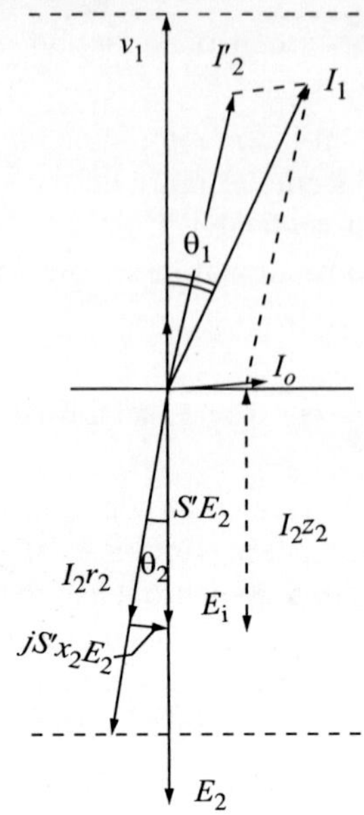

(c)

(i) Speed < h_r,
slip $S' = S$
(ii) P.F. worsened

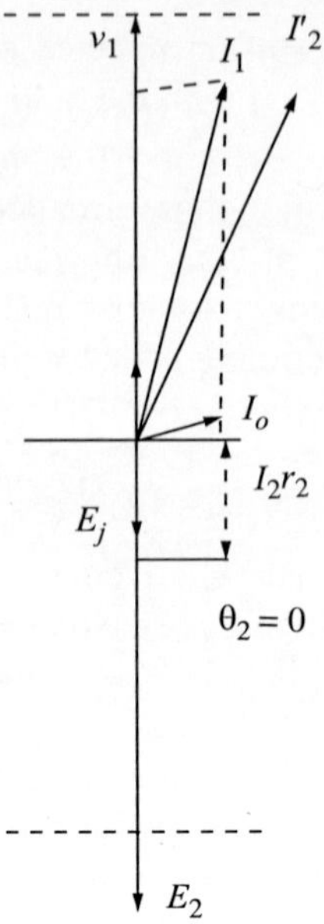

(d)

(i) Speed < h_s,
slip $S' < S$
(ii) F.F. improved

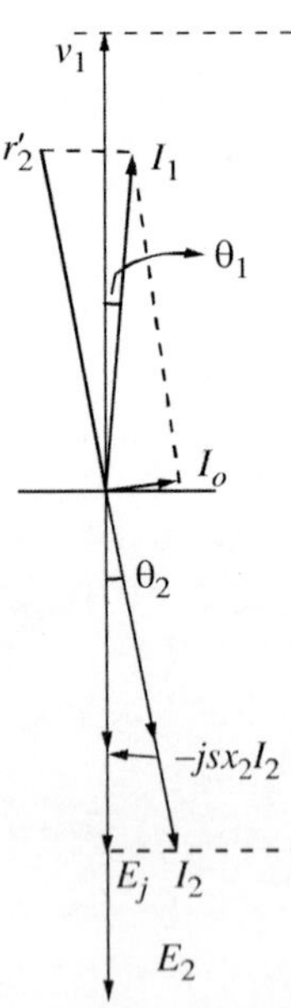

(e)

Speed > h_s
Slip s' is negative
(ii) P.F improved

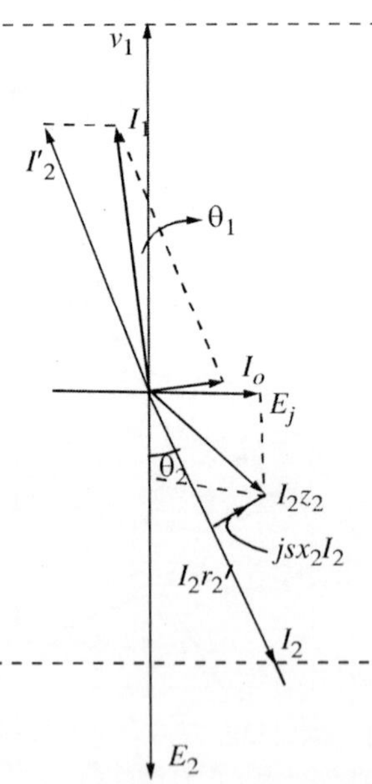

(f)

(i) Speed is effected
(ii) P.F become leading

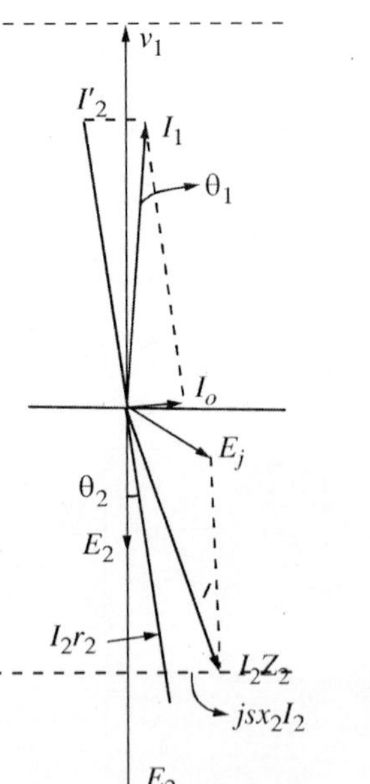

(g) (i) general case
(ii) P.F improved.

Fig 1.15 Effect of EMF injected in the secondary circuit of a polyphase induction motor

The decreased rotor current reduce the electrical torque developed. Since the load-torque is constant and developed electrical torque has decreased, the speed drops and slip emf sE_2 increases as a result of increase in slip emf, the rotor current rises till again electrical torque has become equal to the constant load torque.

Figure 1.16(*b*) illustrate this condition. The physical concept of above happening may be understood more precisely by an analogy. In Fig. 1.10(*a*) A current of 2 Amp. is flowing in the circuit. If an emf E_j = 5 V, is opposition to source voltage of 10 V (equivalent to sE_2) is injected, Fig. 1.12(*b*) the current decreases to 1 Amp. In order to raise the current to its original value of 2A (equivalent to constant $I_2 \cos \theta_2$) the source emf must be increased to 15 V (equivalent to an increase in slip from) S to S'. It can therefore, be calculated that injection of an emf in phase opposition to E_2 result in speeds less than that obtain during normal operation of Fig. 1.10(*a*) the resultant of new rotor emf $s'E_2$ and the injected emf E_j must be I_2Z_2.

If an emf $E_j < I_2Z_2$ is injected in phase with E_2 the rotor current and hence the electrical torque has the effect of increasing the speed of decreasing the slip. As a result of it; slip emf, rotor current and electrical torque decrease till torque balance is obtained. Reference to Fig. 1.16(*c*) show that when E_j = 5 V (less than r_1), aiding the source voltage, is injected, then the source voltage must be decreased to 5 volts, which is equivalent to a decrease in slip. Thus speed above normal speed of Fig. 1.16(*a*) can be obtained by injecting an emf in phase with the rotor voltage E_2. This is illustrated in Fig. 1.16.

If emf $E_j = I_2Z_2$ is injected in phase with E_2, an interesting situation occurs. Reference to Fig. 1.16(*d*) show that when E_j = 10 v the source voltage, equivalent to $s'E_2$ must be zero. In otherwords, with $E_j = I_2Z_2$ and in phase with E_2, the slip $s' = 0$ and rotor now must run at synchronous speed, Fig. 1.16(*d*). Note that the required rotor current is established by injected emf E_j alone.

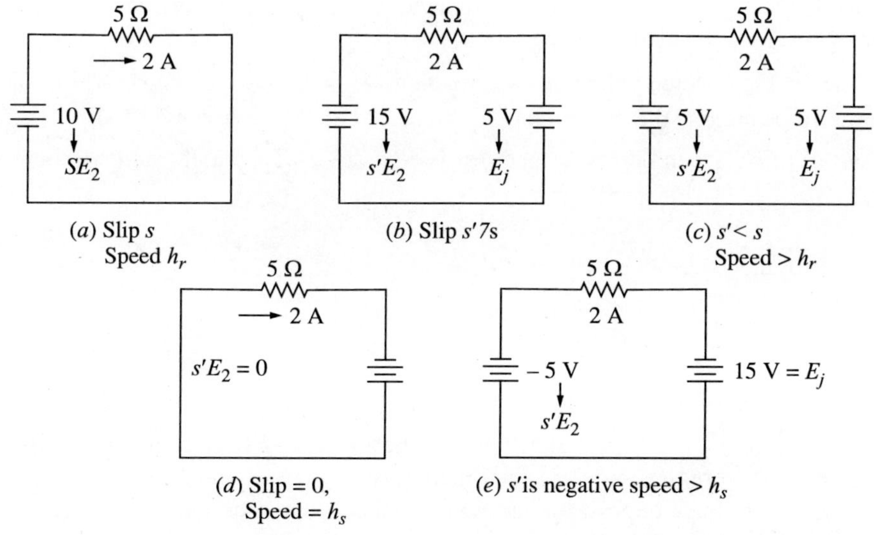

Fig. 1.16 DC Circuit analogy for rotor voltage injection

Now an emf $E_j > I_2Z_2$ is injected in phase with E_2 as shown in Fig. 1.16(*e*). The resultant of $s'E_2$ and E_j must be equal to I_2Z_2 and, therefore, $s'E_2$ and hence s' must be negative for $E_j > I_2Z_2$. The slip s' can be negative only if the rotor speed is above synchronous speed. Reference to Fig. 1.16(*e*) show that $E_j = 15$ V more than *ri* drop of 10 V is injected, the source voltage must be made negative by 5 V, which is equivalent to making $s'E_2$ negative thus the speed above synchronous can be obtained by injecting an emf $E_j > I_2Z_2$ and in phase with E_2.

The effect of injecting E_j at an angle 90° leading the rotor emf E_2 is illustrated in Fig. 1.16(*f*). It is seen from this figure that the phasor orientations of the resultant rotor circuit voltage $(\bar{E}_j + s\bar{E}_2)$ and the rotor current I_2 have changed. The effect of rotor on the stator, or primary side, is to improve the motor *pf* considerably. It may be seen that E_j is either assisting nor opposing E_2; but actually *pf* improvement is accompanied by a change in speed also, as is reversed by example.

Figure 1.16(*g*) illustrates the general case of rotor voltage injection in which E_j is leading E_2 by an angle β and both the speed and power factor are effected.

1.7.1 Rating of the Auxiliary Equipment

The auxiliary equipment inject voltage E_j in the rotor circuit and the rotor current I_2 flow through the electrical circuit of the auxiliary equipment. In view of this, the slip power P_{sp} handled by the auxiliary equipment is given by

$$P_{sp} = m\, E_j\, I_2 \cos\,(E_j, I_2)$$

where m is the number of phases.

If the angle between E_j and I_2 is 90°, then the auxiliary equipment supplies the reactive power only.

1.7.2 Torque-speed Characteristic

The effect of rotor voltage injection on the torque-speed characteristic of an induction motor is now investigated.

When the injected voltage E_j is zero, then $s\bar{E}_2 = \bar{I}\,\bar{Z}_2$ when the voltage is injected in the rotor circuit, the speed and slip change. Let the new value of slip be s'. The result of rotor circuit emf $s'\bar{E}_2$ and $\bar{E}_j$ is equal to $\bar{I}_2\,\bar{Z}_2$, which can be written as

$$s'E_2 + \bar{E}_j = \bar{I}_2\bar{Z}_2$$

with zero injected voltage $s\bar{E}_2 = \bar{I}_2\bar{Z}_2$,

$$s'\bar{E}_2 + \bar{E}_j = s\bar{E}_2$$

or

$$s' = s - \frac{\bar{E}_j}{E_2} \qquad ...(18)$$

If E_j and E_2 are in the same phase, then E_j and E_2 have the same sign and, therefore, s' decreases and the speed increases. In case E_j and E_2 oppose each other then, E_j and E_2 have opposite signs and therefore s' increases and the speed decreases.

The induction motor speed, after emf injection, is given by

$$h_r = n_s\,(1 - s')$$

Substitution of the value of s' from Eq. (19) gives

$$h_r = n_s\left(1 - s + \frac{E_j}{E_2}\right) \quad ...(19)$$

Eq. (19) verifies the conclusions derived from ( ) with E_j and E_2 in the same phase, if $\frac{E_j}{E_2} > s$ or $E_j > sE_2 (= I_2Z_2)$.

The rotor speed h_r becomes more than synchronous speed n_s.

For a given injected voltage, $\frac{E_j}{E_2}$ is constant. At no load, slip s is very small and if it is assumed zero, then the rotor speed at no load is $n_s\left(1 + \frac{\overline{E}_j}{E_2}\right)$ when E_j aids E_2, the no load speed is $n_s\left(1 + \frac{E_j}{E_2}\right)$; when E_j opposes E_2 the no-load speed is $n_s\left[1 - \frac{E_j}{E_2}\right]$ and when $E_j = 0$, the no-load speed is synchronous, this illustrated in Fig. 1.17 the intersection of the torque-speed curves of motor and load gives the operating point and Fig. 1.17 reveals that the operating speed below and above synchronous speed are possible by voltage injection into the rotor circuit of a 3-phase wound-rotor induction motor.

Substituting the value of

$$s = \frac{\overline{I}_2\overline{Z}_2}{E_2},$$

Eq. ( ) may now be written as

$$h_s = h_s\left[1 + \frac{\overline{E}_j}{\overline{E}_2} - \frac{\overline{I}_2\overline{Z}_2}{\overline{E}_2}\right]$$

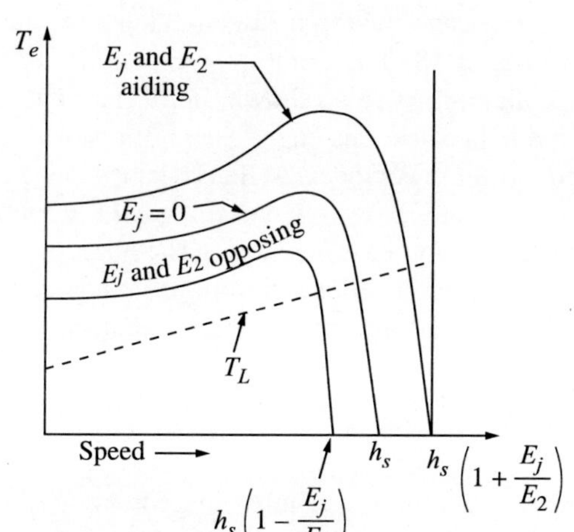

Fig. 1.17 Effect of rotor voltage injection on the torque-speed characteristic of an induction motor

with the inverse in load on the induction motor, rotor current I_2 increases accordingly. Since I_2Z_2 is much smaller in magnitude than E_2, there is only marginal decreases in speed as the load is increased and in view of this, the induction motor torque-speed characteristic of Fig. are obtained.

Eq. (20) may also be re-written as

$$h_r = h_s\left[1+\frac{1}{\bar{E}_2}\left(\bar{E}_j - \bar{I}_2\bar{Z}_2\right)\right] \qquad ....(20)$$

The conclusion drawn from the phasor diagram may also be arrived at from Eq. for example,

(i) if $E_j < I_2Z_2$ and E_j aids E_2 $\qquad h_r < h_s$

(ii) if $E_j = I_2Z_2$ and E_j aids E_2 $\qquad h_r = h_s$

(iii) if $E_j > I_2Z_2$ and E_j aids E_2 $\qquad h_r > h_s$

(iv) if $E_j = 0$; normal rotor speed $= h_s\left[1-\frac{\bar{I}_2\bar{Z}_2}{\bar{E}_2}\right]$ is obtained and

(v) if E_j opposes E_2 then, $h_r = h_s\left[1-\frac{1}{\bar{E}_2}(\bar{E}_j + \bar{I}_2\bar{Z}_2)\right]$ and this speed is less than normal rotor speed of an ordinary induction motor.

1.8 COMPARISON BETWEEN SINGLE-CAGE AND DOUBLE CAGE MOTORS

For comparison purpose, (circle diagram) for double cage induction motor are drawn in the same figure.

The current locus (circle diagram) for double cage induction motor can be drawn by first considering the two cage electrically and magnetically separate and then adding the two current loci.

The bottom cage has high leakage reactance, its circle has, therefore smaller diameter; where as the top cage having a lower leakage reactance has greater circle diameter as shown in Fig. 1.18. The points corresponding to slip s = 1 (*i.e.*, short circuited) on the circle diagram are also shown. For top cage the short circuit point d_2 is near the creast of circle because this cage, has high resistance for bottom cage, the short-circuited point d_z is quit lower, because this cage has lower resistance. The phasor sum of od_2 and od_3 gives the total starting current as od for double-cage motor. Other points on the current locus of double-cage motor. The other points on the current locus of the double-cage motor can be obtained similarly. For comparison purpose, the current locus for a corresponding single-cage induction motor is also drawn in Fig. 1.18.

Now a single cage motor and a double-cage motor of the same rating can be compared as follow:

Starting performance: A double-cage motor develop higher starting current in view of this it is more sutable for direct on line starting. Since effective rotor resistance for a double-cage motor is higher, there is more rotor heating during starting as compared to a single-cage motor.

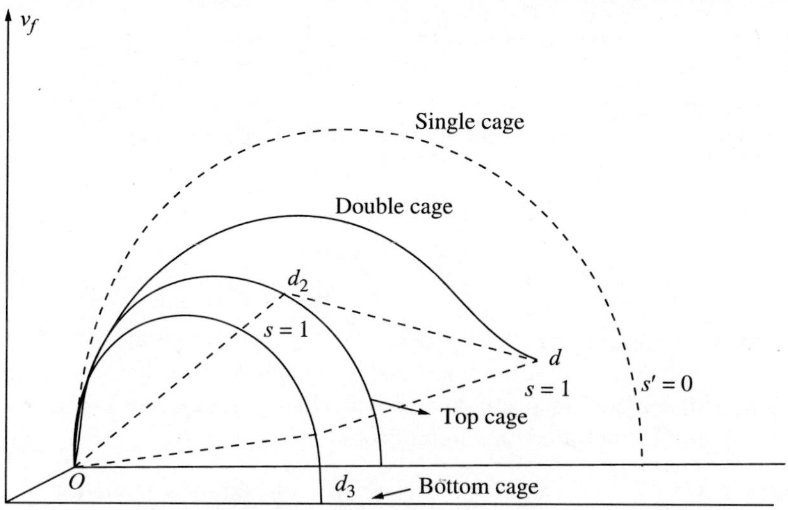

Fig. 1.18 Current locus for a double-cage and single-cage induction motor of same rating

Full load performance: A double cage induction motor has higher effective rotor leakage reactance in view of this, it has lower break down torque and lower full load power factor. Since effective rotor resistance in a double cage motor is higher, it has lower efficiency as compared to a single-cage induction motor. Full load slip of double-cage motor may be lower than the single cage motor.

Torque-slip characteristics: With double cage motor a wide range of torque-slip characteristic can be obtained by a proper choice of top and bottom cage resistance and the slit in between them; this however, is not possible in an ordinary squirrel-cage induction motor.

1.9 CONSTANT TORQUE

If the same torque range is required for both speeds (*i.e.*, power proportional to speed), the windings are best utilized if

$$B_2 \simeq B_1 \text{ or } N_2 kw_2 \, V_1 / N_1 \, kw_1 \, V_2 \simeq P_2/P_1$$

and the ratio of the pull-out torque is then

$$M_1/M_2 \simeq (kw_1/kw_2)^2 \, (P_2/P_1).$$

1.10 CONSTANT POWER

For torque inversely proportional to speed, the ratio $B_1/B_2 = P_1/P_2$ gives the same overload capability on each speed and the same current density in the windings and the phase spreads are inherently, 60° and 120° (dc). The winding factor is $kw = k_d \, k_c$, where for P_1 it is $kw_1 = 0.96 \times 0.71 = 0.68$ and for $P_2 = 2P_1$ it is $kw_2 = 0.83 \times 1.0 = 0.83$. If the phase coil groups are in series and the terminal connection in delta for P_1 and star for P_2, then

$$\frac{N_1}{N_2} = 1, \; \frac{V_1}{V_2} = \frac{1}{\sqrt{3}}, \; \frac{B_1}{B_2} = 0.71, \; \frac{M_1}{M_2} = 1.5$$

$$\frac{I_{m1}}{I_{m2}} = 0.65, \frac{I_{sc1}}{I_{sc2}} = 1.7, \frac{P_1}{P_2} = 3.0$$

So, that constant-torque conditions are roughly achieved for a constant power condition the winding connections could be parallel star/series-delta giving.

$$\frac{N_1}{N_2} = \frac{1}{2}, \frac{V_1}{V_2} = \frac{1}{\sqrt{3}}, \frac{B_1}{B_2} = 0.71, \frac{M_1}{M_2} = 0.68$$

$$\frac{I_{m1}}{I_{m2}} = 0.9, \frac{I_{sc1}}{I_{sc2}} = 1.7, \frac{P_1}{P_2} = 1.4$$

Common connection for $P_2 = 2P_1$ are parallel-star/star and parallel-star/delta which require only six external stator terminals and for which the flux densities do not differ by more than 40 and 20% respectively. On pole changing the stator phase sequence is reversed to presence the same direction of rotation.

Example: A 375 kW, 3.0 kV, 50 Hz 3-phase 10 pole slip-ring induction motor with star/star connection has a rotor resistance $r_2 = 0.39$ Ω/Pb and a total leakage reactance $X_1 = 5.75$ Ω/Ph (both referred to the stator). The stator/rotor effective turns ratio is 4.65 and the full-load slip is 0.022 P.U. Ignoring stator resistance and mechanical loss, estimate the external rotor resistance, the slip power and the efficiency for a speed of one half synchronous when the load has (*i*) constant full-load torque (*ii*) a torque of full-load value of normal speed but is proportional to speed squared. Synchronous speed is 62.8 rad/s per phase, the full load output is 125 hour. The full load torque is 125/(628 × 0.978) = 2.04 kN-M and pull-out torque is 1730^2 (2 × 5.75 × 62.8) = 4.15 kN-M.

Constant torque: Here q = 2.04/4.15 = 0.49 and q^L = 0.24

$$\text{Then } R_2 = 0.50(5.75/0.49)\left[1 \pm \sqrt{(1-0.24)}\right] = 10.97 \text{ (or 0.76) } \Omega$$

Thus $R = R_2 - r_2$ = 9.58 (or 0.37) Ω of which only the former gives a stable operating point. The slip power SP_2 = 0.5(125/0.978) = 63.9 kW/Ph, of which 55.8 kW is dissipated in R. The mechanical output is 125(6.5/0.978) = 63.9 kW, equal to the rotor slip power and with other losses ignored the efficiency is 0.5 P.U. The actual resistance in rotor terms is found by dividing R by the square of the turns ratio; giving $3.58/4.68^2$ = 0.44 Ω.

Torque Proportional to Speed Squared's

At one-half synchronous speed to torque is 2.04 $(0.50/0.978)^2$ = 0.53 kN-M. whence q = 0.53/4.15 = 0.128 and g^L = 0.016, then

R_2 = 44.7 (or 0.18) Ω and R = 44.3(or –0.21) Ω

Showing that the alternative solution is not possible. The slip power is 16.6 kW nearly all dissipated in R_1 and the efficiency is again 0.1 P.U.. In rotor terms R = 2.05 Ω.

1.10.1 Static Slip Power Recovery Control Schemes

These enable slip-ring induction motors to operate of continuously variable speeds efficiently and economically there are two groups.

1.10.2 Mechanical Recovery

Slip power from the slip rings is converted into mechanical power to supplement the output of the main motor.

1.10.3 Electrical Recovery

Slip power is frequency converted and returned to the main supply.

Early examples of the former are there due to Kramer (1906) and of the latter the Scherbiar (1906) and Clymer (1942) systems. All were based on the use of auxiliary machines (dc and ac commulator motors, synchronous motors and convertors. Variable ratio transformers, etc.). In modern variable speed sets, solid state rectifier and inverters can replace most of these.

Simplified examples of mechanical and electrical slip power recovery are shown in below. In both the control of slip is by field control of the dc machine. Decreasing the field reduces the armature emf, allowing increased currents to flow in the slip-ring rotor, which accelerates to a speed determined by the mechanical load. A small difference between the slip-ring and dc armature voltage (referred back through the rectifier) then circulates the load current. In a comparable manner, the speed of the main motor falls it the excilation is raised. In the control characteristics shown, the operating speeds correspond at no load approximately to the intersection of the slip-ring and referred dc armature voltage on load there is a speed drop of 0.05 P.U. or less. The minimum operating (or base) speed is shown in each case as about 0.5 P.U. of synchronous speed.

Electrical recovery applied to a constant-torque load implies that the power input to the induction machine is constant regardless of its speed (but most of the slip power is returned to the supply) with mechanical recovery. The torque contribution of the dc motor reduces the mechanical load taken by the induction motor. The set then takes a constant power input for a drive with torque inversely proportional to speed (i.e. constant mechanical power). In each case the efficiency is reasonably good being lowest of base speed. The mechanical recovery is simpler. while the electrical method is well adapted to high speed drives unsuitable for a direct-coupled dc machine both can be used for driving centrifugal pumps (e.g., for sewage and drainage) and mine ventilating loss.

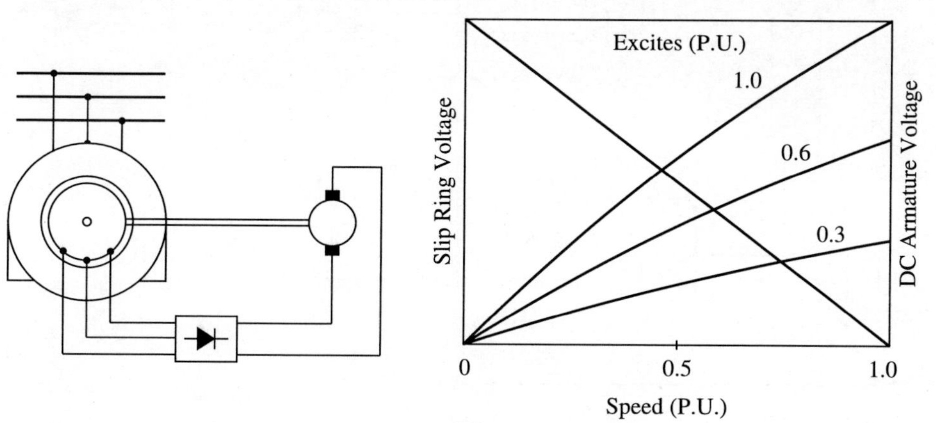

Fig. 1.19

The use of the dc machine is avoided in the modification shown in below where the slip-frequency currents are rectified in a diode bridge network, the unidirectional output current is smoothed and paned on to a 3-ph line-cummulated inverter at a rate depending on the supply voltage, the rectified direct voltage and the thyristor firing angle. The inverted current has a fixed waveform and a constant conduction angle of 2π/3 rad. The onset of conduction with respect to the phase voltage zero is controlled by the firing angle θ. The harmonic content of the inverted current is substantially invariable. In terms of a peak current in of unity, the magnitudes of the fundamental, 5th, 7th, 11th and 13th harmonics are 1.05, 0.24, 0.12, 0.10, and 0.07 as power flow through the rectifier is unidirectional only synchronous speeds are obtainable.

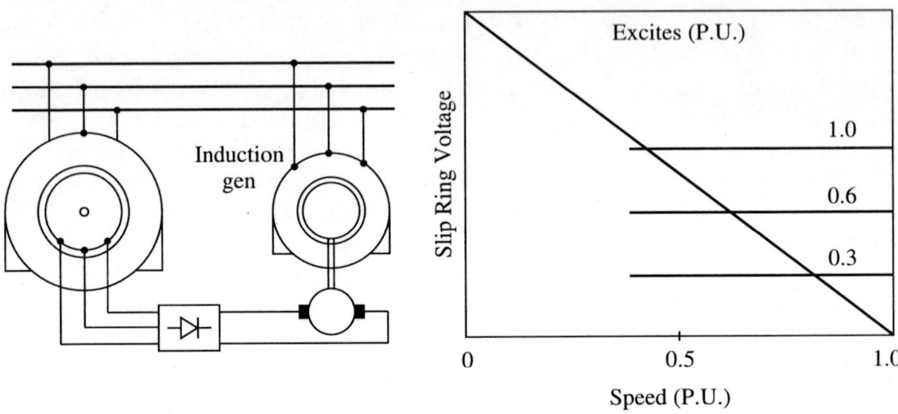

Fig. 1.20 Slip power recovery

In rectifier/inverter recovery, part of the active power input to the motor passes through the rotor back to the supply but the lagging reactive power is absorbed within

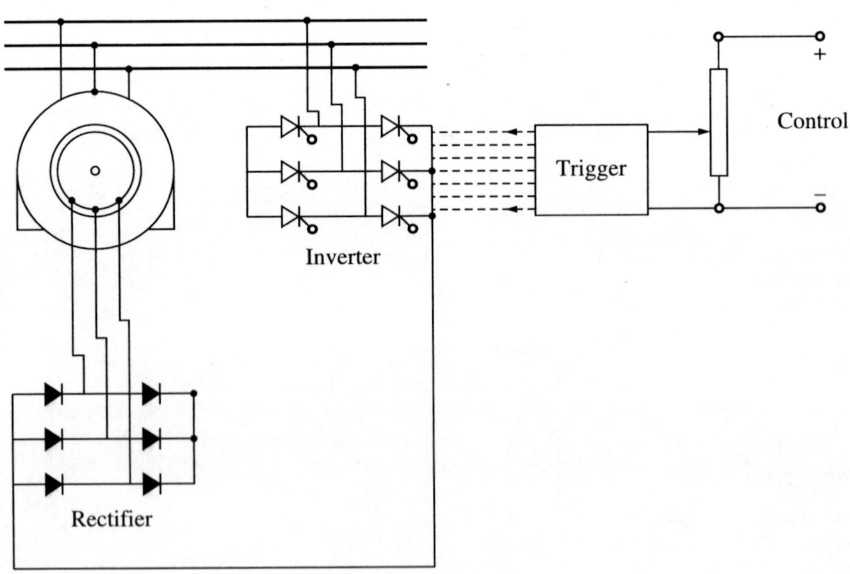

Fig. 1.21 Slip-power recovery

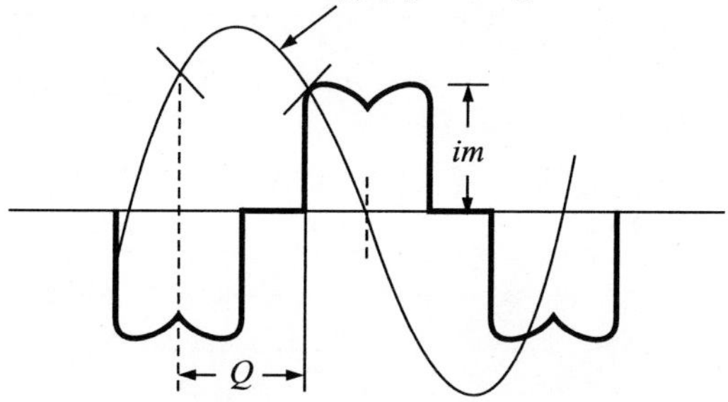

Slip-power recovery

the motor and rectifier. The bridge rectifier is an effect a loss-free lagging reactive powers device a frequency changer and a harmonic current generator, loading the rotor with reactance power to magnetize. The gap at each harmonic frequency, while drawing from the rotor the equivalent in fundamental reactive power. The fundamental of the inverted current leads the phase voltage by an angle equal to the supplement of the firing angle and is therefore a source of leading reactive power (or a sink of lagging reaction power) and the harmonics cause distortion power loss. As regards compensation, the provision of stator capacitors intensifies the harmonics but capacitors across the rotor terminals can markedly decrease the leakage reactance of the motor raise the peak torque.

The usual range of inverter firing angle for wide speed regulation down to standstill is 90°-160° allowing a safe interval for cummulation. The ratio (active/reactive) power of the inverter varies from zero at a 90° firing angle corresponding to maximum speed up to 0.9 lagging for 160°, the low speed condition.

SOLVED NUMERICALS

Example 1.1

A 4 pole, 3 phase induction motor is supplied from 50 Hz supply. Determine its synchronous speed. On full load, its speed is observed to be 1410 rpm. Calculate its full load slip.

Solution: Given values are

$P = 4, \quad f = 50 \text{ Hz}, \quad N = 1410 \text{ rpm}$

$$N_s = \frac{120f}{P} = \frac{120 \times 50}{4} = 1500 \text{ rpm}$$

Full load absolute slip is given by

$$S = \frac{N_s - N}{N_s} = \frac{1500 - 1410}{1500} = 0.06$$

$$\therefore \quad \%S = 0.06 \times 100 = 6\%$$

Example 1.2

A 3-phase, 50 Hz, induction motor has 2 poles. If the slip is 2 percent at a certain load, determine the speed of the motor.

Solution: $f = 50$ Hz, $P = 2$, $S = 2\% = 0.02$

$$N_s = \frac{120f}{P} = \frac{120 \times 50}{2} = 3000 \text{ rpm}$$

$$N = N_s(1 - S)$$

$$N = 3000\ (1 - 0.02)$$

Speed of motor, $N = 2940$ rpm

Example 1.3

A 4 pole, 3 phase, 50 Hz induction motor runs at a speed of 1470 rpm speed. Find the frequency of the induced emf in the rotor under this condition.

Solution: The given values are

$P = 4$, $f = 50$ Hz, $N = 1470$ rpm

$$N_s = \frac{120f}{P} = \frac{120 \times 50}{4} = 1500 \text{ rpm}$$

$$S = \frac{N_s - N}{N_s} = \frac{1500 - 1470}{1500} = 0.02$$

$$\therefore \quad f_r = sf = 0.02 \times 50$$

The frequency of induced emf in the rotor is 1 Hz.

Example 1.4

For a 4-pole, 3-phase, 50 Hz induction motor ratio of stator to rotor turns is 2. On a certain load, its speed is observed to be 1455 rpm when connected to 415 V supply. Calculate

(i) Frequency of rotor emf in running condition.

(ii) Magnitude of induced emf in the rotor at standstill.

(iii) Magnitude of induced emf in the rotor under running condition. Assume star connected stator.

Solution: The given values are,

$$K = \frac{\text{Rotor turns}}{\text{Stator turns}} = \frac{1}{2} = 0.5$$

$P = 4$, $f = 50$ Hz, $N = 1455$ rpm, $E_{1 \text{ line}} = 415$ V

$$N_s = \frac{120f}{P} = \frac{120 \times 50}{4} = 1500 \text{ rpm}$$

For a given load, $N = 1455$ rpm

$$\therefore \quad S = \frac{N_s - N}{N_s} = \frac{1500 - 1455}{1500} = 0.03 \text{ i.e. } 3\%$$

(*i*) $f_r = sf = 0.03 \times 50 = 1.5$ Hz

(*ii*) At standstill, induction motor acts as a transformer so,

$$\frac{E_{2ph}}{E_{1ph}} = \frac{\text{Rotor turns}}{\text{Stator turns}} = K$$

But ratio of stator to rotor turn is given as 2, i.e.

$$\frac{N_1}{N_2} = 2 \quad \therefore \quad \frac{N_2}{N_1} = \frac{1}{2} = K$$

and $E_{1\text{ line}} = 415$ V

$$\therefore \quad E_{1\text{ph}} = \frac{E_1}{\sqrt{3}} = \frac{415}{\sqrt{3}} \qquad [\because \text{ for star connection } E_{1\text{ line}} = \sqrt{3}\, E_{ph}]$$

$$\therefore \quad E_{1\text{ph}} = 239.6 \text{ V}$$

$$\therefore \quad \frac{E_{2ph}}{E_{1ph}} = \frac{1}{2}$$

$$\therefore \quad E_{2\text{ph}} = \frac{1}{2} \times E_{1ph} = \frac{1}{2} \times 239.6 = 119.8 \text{ V}$$

Thus, Rotor induced emf at standstill is 119.8 V.

(*iii*) In running condition

$$E_{2r} = sE_2 = 0.03 \times 119.8 = 3.594 \text{ V}$$

Thus, rotor induced emf at running condition is 3.594 V.

Example 1.5

A 3 phase, 4 pole, 50 Hz, induction motor has slip ring rotor. The rotor winding is star connected with 0.2 Ω of resistance per phase and standstill reactance of 1 Ω per phase. Its open circuit emf between the slip rings is 120 V, when stator is excited by a rated voltage. Its full load speed is 1440 rpm. Find the rotor current and rotor power factor.

(i) At start and (ii) On full load condition.

Solution: The given values are,

$P = 4, \quad f = 50$ Hz, $\quad R = 0.2\ \Omega, \quad X_2 = 1\ \Omega$

Now open circuit emf between slip rings means rotor induced emf on standstill. And between the slip rings means its a line value of E_2, for a star connected rotor.

The open circuit emf is shown in Fig. 1.22.

$$\therefore \quad E_{2\text{line}} = 120 \text{ V, for star } E_{2\text{line}} = \sqrt{3}\, E_{2ph}$$

$$\therefore \quad E_{2\text{ph}} = \frac{E_{2\text{line}}}{\sqrt{3}} = \frac{120}{\sqrt{3}} = 69.28 \text{ V}$$

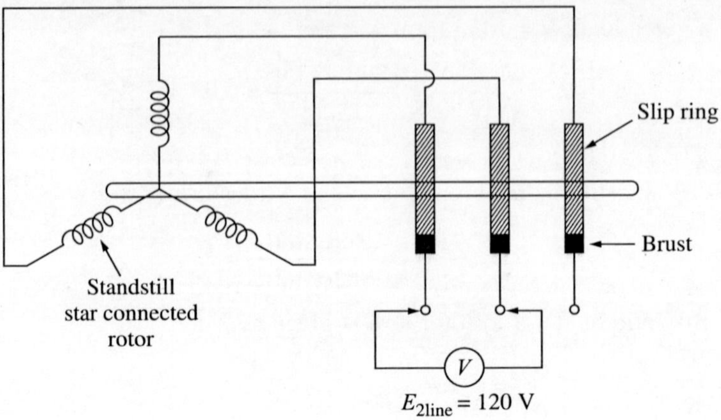

Fig. 1.22

$$N_D = \frac{120f}{P} \times \frac{120 \times 50}{4} = 1500 \text{ rpm}$$

(*i*) At start, $\cos \phi_2 = \dfrac{R_2}{Z_2} = \dfrac{R_2}{\sqrt{R_2^2 + X_2^2}} = \dfrac{0.2}{\sqrt{(0.2)^2 + (1)^2}} = 0.196$ lagging

and $$I_2 = \frac{E_2}{Z_2}$$

$$= \frac{E_2}{\sqrt{R_2^2 + X_2^2}} = \frac{69.28}{\sqrt{(0.2)^2 + (1)^2}} = 67.93 \text{ A per phase}$$

(*ii*) On full load, $N = 1440$ rpm

$$\therefore \quad S = \frac{N_s - N}{N_s} = \frac{1500 - 1440}{1500}$$

$$S = 0.04$$

$$\therefore \quad \cos \phi_{2r} = \frac{R_2}{Z_{2r}} = \frac{R_2}{\sqrt{R_2^2 + (SX_2)^2}} = \frac{0.2}{\sqrt{(0.2)^2 + (0.04 \times 1)^2}}$$

$$= 0.9805 \text{ lagging}$$

and $$I_{2r} = \frac{E_{2r}}{Z_{2r}} = \frac{SE_2}{\sqrt{R_2^2 + (SX_2)^2}} = \frac{0.04 \times 69.28}{\sqrt{(0.2)^2 + (0.04 \times 1)^2}}$$

Rotor current, $I_{2r} = 13.586$ A

Example 1.6

A 3 phase, 400 V, 50 Hz, 4 pole induction motor has star connected stator winding. The rotor resistance and reactance are 0.1 Ω and 1 Ω respectively. The full load speed is 1440 rpm. Calculate the torque developed on full load by the motor. Assume stator to rotor ratio as 2 : 1.

Solution: The given values are,

$P = 4, \quad f = 50$ Hz, $\quad R_2 = 0.1\ \Omega, \quad X_2 = 1\ \Omega, N = 1440$ rpm

$$\frac{\text{Stator turns}}{\text{Rotor turns}} = \frac{2}{1}$$

$$\therefore \quad K = \frac{E_2}{E_1} = \frac{\text{Rotor turns}}{\text{Stator turns}} = \frac{1}{2} = 0.5$$

$$N_s = \frac{120}{P} = \frac{120 \times 50}{4} = 1500 \text{ rpm}$$

$$E_{1\text{ line}} = 400 \text{ V} \qquad \text{stator line voltage given}$$

$$\therefore \quad E_{1\text{ ph}} = \frac{E_{1\text{line}}}{\sqrt{3}} = \frac{400}{\sqrt{3}} = 230.94 \text{ V}$$

But
$$\frac{E_{2\text{ph}}}{E_{1\text{ph}}} = 0.5 = K$$

$$\therefore \quad E_{2\text{ ph}} = 0.5 \times 230.94 = 115.47 \text{ V}$$

$$\text{Full load slip } S = \frac{N_s - N}{N_s} = \frac{1500 - 1440}{1500} = 0.04$$

$$n_s = \text{Synchronous speed in rps}$$

$$n_s = \frac{N_s}{60} = \frac{1500}{60} = 25 \text{ rps}$$

$$T = \frac{3}{2\pi n_s} \frac{sE_2^2 R_2}{R_2^2 + (sX_2)^2}$$

$$= \frac{3}{2\pi \times 25} \times \frac{0.04 \times (115.47)^2 \times 0.1}{[(0.1)^2 + (0.04 \times 1)^2]}$$

$$T = 87.81 \text{ N-m}$$

Example 1.7

A 400 V, 4 pole, 3 phase, 50 Hz star connected induction motor has a rotor resistance and reactance per phase equal to 0.01 Ω and 0.1 Ω respectively. Determine (i) starting torque (ii) slip at which maximum torque will occur (iii) speed at which maximum torque will occur. (iv) Maximum torque (v) Full load torque if full load slip is 4%. Assume ratio of stator to rotor turn as 4.

Solution: The given values are

$$P = 4, \quad f = 50 \text{ Hz.} \quad \frac{\text{Stator turns}}{\text{Rotor turns}} = 4$$

$R_2 = 0.01\ \Omega, \quad X_2 = 0.1\ \Omega$

$$E_{1\text{ line}} = \text{Stator line voltage} = 400 \text{ V}$$

$$E_{1\text{ ph}} = \frac{E_{1\text{line}}}{\sqrt{3}} = \frac{400}{\sqrt{3}} = 230.94 \text{ V}$$

$$K = \frac{E_{2\,\text{ph}}}{E_{1\,\text{ph}}} = \frac{\text{Rotor turns}}{\text{Stator turns}} = \frac{1}{4}$$

$$\therefore \quad E_2 = \frac{1}{4} \times E_{1\,\text{ph}} = \frac{230.94}{4} = 57.735 \text{ V}$$

$$N_s = \frac{120\,f}{P} = \frac{120 \times 50}{4} = 1500 \text{ rpm}$$

(*i*) At start $S = 1$

$$\therefore \quad T_{\text{st}} = \frac{KE_2^2 R_2}{R_2^2 + X_2^2} \quad \text{where } K = \frac{3}{2\pi n_s}$$

$$n_s = \frac{N_s}{60} = \frac{1500}{60} = 25 \text{ rps}$$

$$\therefore \quad K = \frac{3}{2\pi \times 25} = 0.01909$$

$$\therefore \quad T_{\text{st}} = \frac{0.01909 \times (57.735)^2 \times 0.01}{(0.01)^2 + (0.1)^2}$$

$$T_{\text{st}} = 63.031 \text{ N-m}$$

(*ii*) Slip at which maximum torque occurs is

$$S_m = \frac{R_2}{X_2} = \frac{0.01}{0.1} = 0.1$$

$$\%S_m = 0.1 \times 100 = 10\%$$

(*iii*) Speed at which maximum torque occurs is speed corresponding to S_m.

$$N = N_s\,(1 - S_m) = 1500\,(1 - 0.1) = 1350 \text{ rpm}$$

(*iv*) The maximum torque is,

$$T_m = \frac{KE_2^2}{2X_2} = \frac{0.01909 \times (57.735)^2}{2 \times 0.1} = 318.16 \text{ N-m}$$

(*v*) Full load slip, $s_f = 0.04$ as $\%s_f = 4\%$

$$\therefore \quad T_{f.l} = \frac{K\,s_f\,E_2^2 R_2}{R_2^2 + (s_f X_2)^2} = \frac{0.01909 \times 0.04 \times (57.735)^2 \times 0.01}{(0.01)^2 + (0.04 \times 0.1)^2}$$

$$T_{f.l.} = 219.52 \text{ N-m}$$

Example 1.8

A 24 pole, 50 Hz, star connected induction motor has rotor resistance of 0.016 Ω per phase and rotor reactance of 0.265 Ω per phase at standstill. It is achieving its full load torque at a speed of 247 rpm. Calculate the ratio of

(*i*) *Full load torque to maximum torque*

(*ii*) *Starting torque to maximum torque*

Solution: The given values are

$P = 24, f = 50$ Hz, $R_2 = 0.016\ \Omega$, $X_2 = 0.265\ \Omega$

$$N = 247 \text{ rpm}$$

$$N_s = \frac{120f}{P} = \frac{120 \times 50}{24} = 250 \text{ rpm}$$

$$s_f = \frac{N_s - N}{N_s} = \frac{250 - 247}{250} = 0.012 = \text{Full load slip}$$

$$S_m = \frac{R_2}{X_2} = \frac{0.016}{0.265} = 0.06037$$

(i) $$\frac{T_{F.L.}}{T_m} = \frac{2S_m.S_f}{S_m^2 + S_f^2} = \frac{2 \times 0.06037 \times 0.012}{(0.06037)^2 + (0.012)^2} = 0.3824$$

(ii) $$\frac{T_{st}}{T_m} = \frac{2S_m}{1 + S_m^2} = \frac{2 \times 0.06037}{1 + (0.06037)^2} = 0.1203$$

Example 1.9

Rotor resistance and standstill per phase of a 3 phase induction motor are 0.04 Ω and 0.2 Ω respectively. What should be the external resistance required at start in rotor circuit to obtain

(i) Maximum torque at start

(ii) 50% of maximum torque at start.

Solution: $R_2 = 0.04\ \Omega$, $X_2 = 0.2\ \Omega$

(*i*) For $T_m = T_{st}$, $S_m = \dfrac{R_2'}{X_2} = 1$

$\therefore\ R_2' = X_2 = 0.2\ \Omega$

Let R_{ex} = External resistance required in rotor.

$$R_2' = R_2 + R_{ex}$$

$$\therefore \quad R_{ex} = R_2' - R_2 = 0.2 - 0.04 = 0.16 \text{ per phase}$$

(*ii*) For $T_{st} = \dfrac{1}{2} T_m$

Now, $$T_m = \frac{kE_2^2}{2X_2} \text{ and}$$

$$T_{st} = \frac{k\,E_2^2 R_2}{R_2^2 + X_2^2}$$

But at start, external resistance R_{ex} is added, so new value of rotor resistance is say R_2'.

$$R_2' = R_2 + R_{ex}$$

$$\therefore \qquad T_{st} = \frac{k\,E_2^2 R_2'}{R_2'^2 + X_2^2} \text{ with added resistance}$$

$$\text{But} \qquad T_{st} = \frac{1}{2} T_m \text{ required}$$

Substituting expressions of T_{st} and T_m, we get

$$\therefore \qquad \frac{k\,E_2^2 R_2'}{(R_2')^2 + X_2^2} = \frac{1}{2} \cdot \frac{k\,E_2^2}{2X_2}$$

$$\therefore \qquad 4R_2'X_2 = (R_2')^2 + (X_2)^2$$

$$\therefore \qquad (R_2')^2 - 4 \times 0.2 \times R_2' + (0.2)^2 = 0$$

$$\therefore \qquad (R_2')^2 - 0.8R_2' + 0.04 = 0$$

$$\therefore \qquad R_2' = \frac{0.8 \pm \sqrt{(0.8)^2 - 4 \times 0.04}}{2}$$

$$\therefore \qquad R_2' = 0.0535\ \Omega,\ 0.7464\ \Omega.$$

But R_2' cannot be greates than X_2, hence

$$\therefore \qquad R_2' = 0.0535 = R_2 + R_{ex}$$

$$\therefore \qquad R_{ex} = R_2' - R_2 = 0.0535 - 0.04$$

$$R_{ex} = 0.0135\ \Omega \text{ per phase}$$

This much resistance is required in the rotor externally to obtain $T_{st} = \frac{1}{2} T_m$.

Example 1.10

The useful torque of a 3 phase, 50 Hz, 8 pole induction motor is 190 N-m. The rotor frequency is 1.5 Hz. Calculate the rotor copper losses if mechanical losses are 700 watts.

Solution: T_{sh} = 190 N-m, f_r = 1.5 Hz

Mechanical loss = 700 W

$$\text{Now} \qquad f_r = sf$$

$$\therefore \qquad 1.5 = s \times 50$$

$$\therefore \qquad s = \frac{1.5}{50} = 0.03$$

$$\text{Now} \qquad P_{out} = T_{sh} \times w$$

$$N = N_s(1 - s)$$

$$\text{and} \qquad N_s = \frac{120\,f}{P} = \frac{120 \times 50}{8} = 750 \text{ rpm}$$

$$\therefore \qquad N = 750\,(1 - 0.03) = 727.5 \text{ rpm}$$

$$\therefore \qquad P_{out} = T_{sh} \times \frac{2\pi N}{60} = 190 \times 2\pi \times \frac{727.5}{60}$$

$$P_{out} = 14474.88 \text{ W}$$

Now, P_m – Mechanical losses = P_{out}

$$\therefore \quad P_m - 700 = 14474.88$$

$$\therefore \quad P_m = 700 + 14474.88$$

$$P_m = 15174.88 \text{ W}$$

from $P_2 : P_c : P_m$ is $1 : s : 1 - s$

$$\therefore \quad \frac{P_c}{P_m} = \frac{s}{1-s}$$

$$P_c = P_m \times \frac{s}{1-s} = 15174.88 \times \frac{0.03}{(1-0.03)}$$

Rotor copper loss, $P_c = 469.326$ W

Example 1.11

The full load power input to 4 pole, 50 Hz, three phase induction motor is 50 KW, running at 1440 rpm. Calculate its full load efficiency if stator losses are 1000 W and frictional losses are 650 W.

Solution: Given values are

$P = 4, f = 50, P_{in} = 50$ KW, $N = 1440$ rpm

Stator losses = 1000 W

Frictional losses = 650 W

$$N_s = \frac{120f}{P} = \frac{120 \times 50}{4} = 1500 \text{ rpm}$$

$$S = \frac{N_s - N}{N_s} = \frac{1500 - 1440}{1500} = 0.04$$

$$P_2 = \text{Rotor input} = \text{Pin} - \text{Stator losses}$$

$$P_2 = 50 \times 10^3 - 1000 = 49 \times 10^3 \text{ W.}$$

Now $P_2 : P_c : P_m$ is $1 : S : 1 - S$

$$\therefore \quad \frac{P_2}{P_c} = \frac{1}{S}$$

$$\therefore \quad P_c = S \times P_2 = 0.04 \times 49 \times 10^3 = 1960 \text{ W}$$

$$\therefore \quad P_m = P_2 - P_c = 49 \times 10^3 - 1960 = 47040 \text{ W}$$

Now $\quad P_{out} = P_m$ – Frictional losses = 47040 – 650

$$P_{out} = 46390 \text{ W}$$

$$\therefore \quad \%\eta = \frac{P_{out}}{P_{in}} \times 100$$

$$\%\eta = \frac{46390}{50 \times 10^3} \times 100 = 92.78\%.$$

As slip is full load, efficiency is full load efficiency.

Example 1.12

A 3 phase, 4 pole, 50 Hz, star connected induction motor running on full load develops a useful torque of 300 N-m. The rotor emf is completing 120 cycles per minutes. If torque lost in friction is 50 N-m, calculate:

(i) Slip
(ii) Net output power
(iii) Rotor copper loss per phase
(iv) Rotor efficiency
(v) Rotor resistance per phase if rotor current is 60 A in running condition.

Solution: The given values are

$P = 4, f = 50$ Hz, $T_{sh} = 300$ N-m, $T_{rest} = 50$ N-m

Rotor frequency = 120 cycles/min = $\frac{120}{60}$ cycles/sec *i.e.* Hz.

i.e. $f_r = 2$ Hz

(*i*) $f_r = sf$

$$\therefore \quad s = \frac{f_r}{f} = \frac{2}{50} = 0.04 \text{ i.e. } 4\%$$

(*ii*) $P_{out} = T_{sh} \times \omega = T_{sh} \times \frac{2\pi N}{60}$

Now $N = N_s(1 - s)$ at slip $s = 0.04$

$$N_s = \frac{120\, f}{P}$$

$$N_s = \frac{120 \times 50}{4} = 1500 \text{ rpm}$$

$$\therefore \quad N_s(1 - s) = 1500\,(1 - 0.04)$$

$$N = 1440 \text{ rpm}$$

$$\therefore \quad P_{out} = 300 \times \frac{2\pi \times 1440}{60}$$

$$P_{out} = 45.238 \text{ kW.}$$

Remember that T_{sh} is net output torque available to load at shaft.

(*iii*) $T_{lost} = 50$ N-m in friction.

$$\therefore \quad \text{Frictional loss} = T_{lost} \times w$$

$$= T_{lost} \times \frac{2\pi N}{60}$$

$$= 50 \times \frac{2\pi \times 1440}{60}$$

$$= 7539.822 \text{ W}$$

Now $P_{out} = P_m -$ Frictional loss

$$\therefore \quad P_m = P_{out} + \text{Frictional loss}$$

$$P_m = 45.2389 \times 10^3 + 7539.822$$

$$P_m = 52.77872 \text{ kW}$$

We know $P_2 : P_c : P_m$ is $1 : s : 1 - s$

$$\therefore \quad \frac{P_c}{P_m} = \frac{s}{1-s}$$

$$\therefore \quad P_c = P_m\left(\frac{s}{1-s}\right) = 52.77872 \times 10^3 \times \left(\frac{0.04}{1-0.04}\right)$$

Total Cu loss, $P_c = 2199.1134$ W

$$\therefore \text{ Rotor copper loss per phase} = \frac{P_c}{3} = \frac{2199.1134}{3} = 733.0378 \text{ W}$$

(*iv*) Rotor efficiency $= \dfrac{\text{Rotor output } P_m}{\text{Rotor input } P_2} \times 100$

Now $$P_2 = \frac{P_c}{s} = \frac{2199.1139}{0.04} = 54977.8358 \text{ W}$$

$$\therefore \quad \% \text{ Rotor } \eta = \frac{52778.72}{54977.83} \times 100 = 96\%$$

(*v*) $I_{2r} = 60$ A given per phase

Now rotor copper loss/ph $= I_{2r}^2 \times R_2$

$$\therefore \quad 733.0378 = (60)^2 \times R_2$$

$$\therefore \quad R_2 = 0.2036 \text{ W/ph}$$

Example 1.13

A 25 kW, 4 pole, 3 phase, 50 Hz induction motor is running at 1410 rpm supplying full load. The mechanical losses are 850 W and stator losses are 1.7 times rotor copper losses on full load. Calculate,

(*i*) *Gross mechanical power developed*

(*ii*) *Rotor copper losses*

(*iii*) *The value of rotor resistance per phase if rotor current on full load per phase is 65 A.*

(*iv*) *The full load efficiency.*

Solution: The given rating 25 kW of the motor is full load shaft output.

$$\therefore \quad P_{out} = 25 \text{ kW}$$

$$N_s = \frac{120 f}{P} = \frac{120 \times 50}{4} = 1500 \text{ rpm}$$

$$N = 1410 \text{ rpm}$$

$$\therefore \quad S = \frac{N_s - N}{N_s} = \frac{1500 - 1410}{1500} = 0.06$$

Now P_m – Mechanical losses = P_{out}

$\therefore \quad P_m - 850 = 25 \times 10^3$

(*i*) P_m = Gross mechanical power

$P_m = 25 \times 10^3 + 850$

$P_m = 25850$ W

(*ii*) Now $P_2 : P_c : P_m$ is $1 : s : 1 - s$

$$\therefore \quad \frac{P_c}{P_m} = \frac{s}{1-s}$$

$$\therefore \quad P_c = 25850 \times \frac{0.06}{(1-0.06)}$$

$$P_c = 1650 \text{ W}$$

These are total rotor copper loss.

(*iii*) $I_{2r} = 65$ A

Now $\quad P_c = 3 \times I_{2r}^2 \times R_2$

$$\therefore \quad R_2 = \frac{P_c}{3 \times (I_{2r})^2} = \frac{1650}{3 \times (65)^2}$$

$$R_2 = 0.13 \ \Omega \text{ per phase}$$

(*iv*) Stator losses = $1.7 \times P_c = 1.7 \times 1650 = 2805$ W

Now, $\quad P_c = s \times P_2$

$$P_2 = \frac{P_c}{s} = \frac{1650}{0.06} = 27500 \text{ W}$$

$$P_2 = 27500 \text{ W}$$

$\therefore \quad P_{in}$ – stator losses = P_2

$$P_{in} = P_2 + \text{stator losses}$$

$$P_{in} = (27500 + 2805) \text{ W}$$

$$P_{in} = 30305 \text{ W}$$

$$\therefore \quad \%\eta = \frac{P_{out}}{P_{in}} \times 100 = \frac{25000}{30305} \times 100$$

$$\%\eta = 82.49\%$$

Example 1.14

While delivering an useful power of 24 kW to the full load, a 3 phase, 50 Hz, 8 pole induction motor draws a line current of 57 A. It runs at a speed of 720 rpm and is connected to 415 V supply. The pf of the motor is observed to be 0.707 lagging. Stator resistance per phase is 0.1 Ω. Mechanical losses are 1000 watts. Calculate,

(*i*) *Shaft torque*

(*ii*) *Gross torque developed*

(*iii*) *Rotor copper losses*

(iv) *Stator copper losses*

(v) *Stator iron losses*

(vi) *Overall efficiency.*

Assume star connected stator winding.

Solution: $P_{out} = 24$ kW, $I_L = 57$ A

$P = 8$, $N = 720$ rpm, $f = 50$ Hz

$$V_L = 415 \text{ V}$$

$$\cos \phi = 0.707$$

$$N_s = \frac{120 f}{P} = \frac{120 \times 50}{8} = 750 \text{ rpm}$$

$$\therefore \quad S = \frac{N_s - N}{N_s} = \frac{750 - 720}{8} = 0.04$$

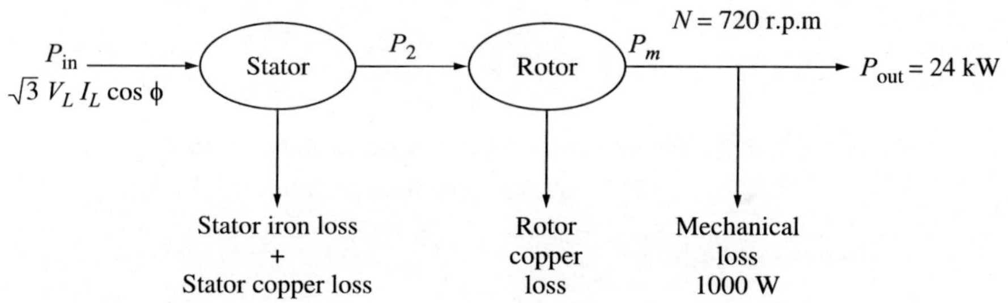

Fig. 1.23

P_m – Mechanical loss = P_{out}

$$\therefore \quad P_m = P_{out} + \text{Mechanical loss}$$

$$P_m = 24 \times 10^3 + 1000$$

$$P_{in} = 25000 \text{ W}$$

For rotor $P_2 : P_c : P_m$ is $1 : s : 1 - s$

$$\therefore \quad \frac{P_c}{P_m} = \frac{s}{1-s}$$

$$\therefore \quad P_c = P_m \times \left(\frac{s}{1-s}\right)$$

$$P_c = 25000 \times \frac{0.04}{(1-0.04)}$$

$$P_c = 1041 = 66 \text{ W}$$

(*i*) Short torque $T_{sh} = \dfrac{P_{out}}{\omega} = \dfrac{P_{out}}{\left(\dfrac{2\pi N}{60}\right)}$

$$= \frac{24 \times 10^3}{\left(\dfrac{2\pi \times 720}{60}\right)} = 318.309 \text{ N-m}$$

(*ii*) Gross torque $T = \frac{P_m}{\omega} = \frac{P_m}{\left(\frac{2\pi N}{60}\right)}$

$$= \frac{25 \times 10^3}{\left(\frac{2\pi \times 720}{60}\right)} = 331.572 \text{ N-m}$$

(*iii*) Rotor copper loss = 1041.66 W

Now $\frac{P_2}{P_c} = \frac{1}{s}$

$\therefore \quad P_2 = \frac{1}{s} \times P_c = \frac{1}{0.04} \times 1041.66 = 26041.5 \text{ W}$

And net input $P_{in} = \sqrt{3} V_L I_L \cos\phi$

$$P_{in} = \sqrt{3} \times 415 \times 57 \times 0.707$$

$$P_{in} = 28966.96 \text{ W}$$

Stator current per phase $= I_L = 57$ A (as star connected) $= I_s$

R_s = Stator resistance per phase = 0.1 W

Stator copper losses $= 3 \times I_s^2 \times R_s$

$= 3 \times (57)^2 \times 0.1 = 974.7$ W

Now P_{in} – stator losses = P_2

$\therefore$ 28966.96 – Stator losses = 26041.5

Stator losses = 2925.46 W

But stator losses = stator iron loss + stator copper

$\therefore$ 2925.46 = Stator iron losses + 974.7

$\therefore$ Stator iron losses = 1950.76 W

(*iv*) Stator copper losses = 974.7 W

(*v*) Stator iron losses = 1950.76 W

(*vi*) $\%\eta = \frac{P_{out}}{P_{in}} \times 100 = \frac{24 \times 10^3}{28966.96} \times 100$

$\%\eta = 82.85\%$

Example 1.15

The following data refers to a 12 pole, 420 V, 50 Hz, three phase mesh connected induction motor.

r_1 = 2.95 W, x_1 = 6.82 W, r_2' = 2.08 W, x_2' = 4.11 Ω per phase.

On no load, the line value of magnetising current is 6.7 A and the total core loss is 269 W. Determine the power factor, input current, equivalent rotor current and torque developed by the motor at a slip of 3% using exact equivalent circuit. Determine the maximum torque developed and the corresponding speed.

Solution: On no load, the equivalent circuit is as shown in Fig. 1.24.

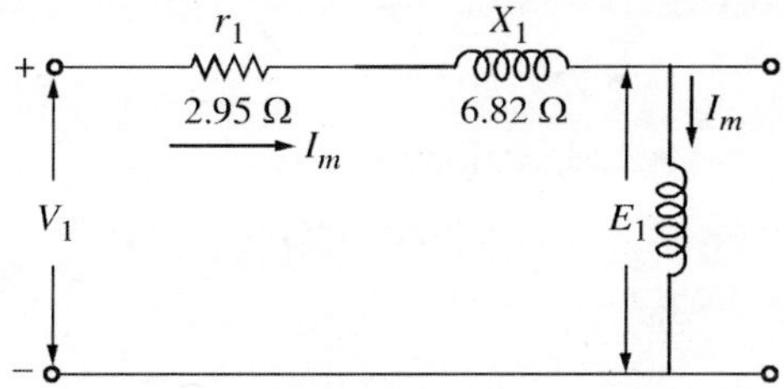

Fig. 1.24

$$I_{m(\text{line})} = 6.7 \text{ A}$$

As motor is delta connected

$$I_m = \frac{I_{m(\text{line})}}{\sqrt{3}} = 3.868 \text{ A}$$

$$\bar{V}_1 = \bar{E}_1 + \bar{I}_m(r_1 + jX_1)$$

Now I_m lags E_1 by 90° and taking E_1 as a reference

$$\bar{I}_m = 3.868\angle{-90°}\ (2.95 + j6.82)$$

$$V_1 = E_1 + 3.868\angle{90°} \times 7.4306\angle{66.609°}$$

$$V_1 = E_1 + 28.7415\angle{-23.391°} = (E_1 + 26.379) - j\ 11.4104$$

Equating magnitude of both side

$$\therefore \quad V_1^2 = (E_1 + 26.379)^2 + (11.4104)^2 \ ...\ V_1 = 420 \text{ V}$$

$$\therefore \quad (420)^2 - (11.4104)^2 = E_1^2 + 52.758\, E_1 + (26.379)^2$$

Solving $E_1 = 393.46$ V Neglecting –ve values

$$\therefore \quad X_0 = \frac{E_1}{I_m} = \frac{393.46}{3.868} = 101.721\ \Omega$$

The exact equivalent circuit is shown in Fig. 1.25.

$$Z_{eq} = jX_0 \left\| \left(\frac{r_2'}{s} + jX_2' \right) \right\| = (j101.721) \,\|\, (69.33 + j4.11)$$

$$Z_{eq} = \frac{101.721\angle{90°} \times 69.4517\angle{3.39°}}{0 + j101.721 + j4.11 + 69.33} = \frac{7064.6963\angle{93.39°}}{126.3518\angle{56.771°}}$$

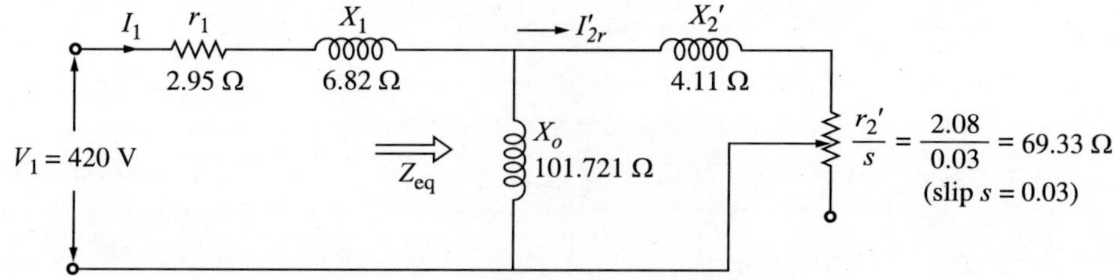

Fig. 1.25

$$\therefore \quad Z_{in} = r_1 + jX_1 + Z_{eq} = 2.95 + j\,6.82 + 44.8172 + j\,33.308$$

$$= 47.7672 + j\,40.128\ \Omega = 62.385\angle 40.04^\circ\ \Omega$$

$$\therefore \quad I_1 = \frac{V_1}{Z_{in}} = \frac{420\angle 0^\circ}{62.385\angle 40.04^\circ} = 6.732\angle 40.04^\circ\ \Omega \text{ (input current)}$$

$$\therefore \quad \cos\phi = \cos(-40.04^\circ) = 0.7655 \text{ lagging power factor}$$

The equivalent rotor current I'_{2r} can be obtained by using current distribution rule in parallel circuit as

$$I'_{2r} = I_1 \frac{jX_0}{jX_0 + r'_2 + jX'_2} = \frac{6.732\angle -40.04^\circ \times 101.721\angle 90^\circ}{126.518\angle 56.771^\circ}$$

$$= 5.413\angle -6.811^\circ\ \text{A}$$

The torque developed is,

$$T = \frac{9.55 \times 3 \times (I'_{2r})^2 r'_2}{sN_s} \quad \text{where } N_s = \frac{120f}{P} = 500 \text{ rpm}$$

$$= \frac{9.55 \times 3 \times (5.413)^3 \times 2.08}{0.03 \times 500} = 116.405 \text{ N-m}$$

To find the maximum torque Tm. first calculate $R_{THQ}\ X_{TH}$.

$$Z_{TH} = R_{TH} + jX_{TH} = (r_1 + j \times 1) \parallel (j \times 0)$$

$$= (2.95 + j6.82) \parallel (j101.721)$$

$$= \frac{7.4306\ \angle 66.6^\circ \times 101.721\ \angle 90^\circ}{2.95 + j6.82 + j101.721}$$

$$= \frac{7.55.848\ \angle 156.6^\circ}{108.581\ \angle 88.4431} = 6.9611\ \angle 68.157^\circ\ \Omega$$

$$Z_{TH} = 2.59 + j6.461\ \Omega$$

$$\therefore \quad R_{TH} = 2.59\ \Omega \text{ and } X_{TH} = 6.461\ \Omega$$

$$V_{TH} = \frac{V_1(jX_0)}{r_1 + j(X_{TH} + X_0)} = \frac{420 \times 101.721\ \angle 90^\circ}{2.95 + j(6.461 + 101.721)}$$

$$V_{TH} = \frac{42722.8^2\ \angle 90^\circ}{108.222\ \angle 88.44^\circ} = 394.77\ \angle 1.56^\circ V$$

$$\therefore \quad T_m = \frac{3}{w_s} \times \frac{0.5\,V_{TH}^2}{R_{TH} + \sqrt{R_{TH}^2 + (X_{TH} + X'_2)^2}}$$

$$T_m = \frac{3}{\left(\dfrac{2\pi \times 500}{60}\right)} \times \frac{0.5 \times (394.77)^2}{2.59 + \sqrt{(2.59)^2 + (6.461 + 4.11)}}$$

$$T_m = 331.35 \text{ N-m}$$

$$S_m = \frac{r_2'}{\sqrt{R_{TH}^2 + (X_{TH} + X_2')^2}} \text{ ... Slip at maximum torque}$$

$$S_m = \frac{2.08}{\sqrt{(2.59)^2 + (6.461 + 4.11)^2}} = 0.1911 \text{ } i.e., \text{ } 19.11\%$$

$\therefore$ N_m = Speed at maximum torque = $N_S(1 - Sm)$

N_m = 500 (1 – 0.1911)

N_m = 404.4 r.p.m.

Example 1.16

A 440 V, 3 phase, 8 pole, 50 Hz, 40 kW, star connected three phase induction motor has the following parameters.

Stator resistance = 0.1 Ω, Stator reactance = 0.4 Ω

Equivalent rotor resistance referred to stator = 0.15 Ω

Equivalent rotor reactance referred to stator = 0.44 Ω

The stator core loss is 1250 W while mechanical loss is 1000 W. It draws a no load current of 20A awt a p.f. of 0.09 lagging, while running at a speed of 727.5 r.p.m calculate.

(i) Input line current and p.f.

(ii) Torque developed

(iii) Output power and

(iv) Efficiency

Use approximate equivalent circuit.

Solution: The approximate equivalent circuit is shown in Fig. 1.26.

$$N_s = \frac{120f}{P} = \frac{120 \times 50}{8} = 750 \text{ r.p.m}$$

$$S = \frac{N_s - N}{N_s} = \frac{750 - 727.5}{750} = 0.03$$

$$R_L' = \frac{R_2'(1-S)}{S} = \frac{0.15(1-0.03)}{0.03} = 4.85\ \Omega$$

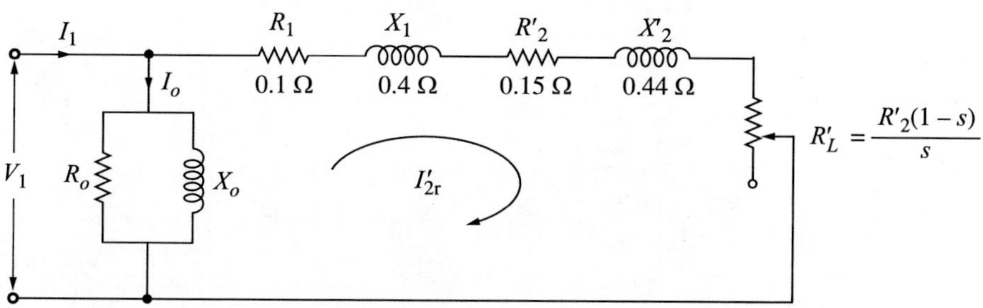

Fig. 1.26

$$V_1 = \frac{440}{\sqrt{3}} = 254.034 \text{ V per phase}$$

$$I'_{2r} = \frac{V_1}{(R_{1e} + Ri) + j(X_{1e})}$$

where $R_{1e} = R_1 + R'_2 = 0.1 + 0.15 = 0.25\ \Omega$

$$X_{1e} = X_1 + X'_2 = 0.4 + 0.44 = 0.84\ \Omega$$

$$\therefore \quad I'_{2r} = \frac{(254.034)}{(0.25 + 4.85) + j(0.84)} = \frac{254.034}{5.1687 \angle +9.3529^\circ}$$

$$= 49.1484 \angle -9.3529^\circ \text{ A using } R \to P$$

$$= 48.495 - j7.9873 \text{ A using } P \to R$$

$$I_0 = 20 \text{ A at } \cos\phi_0 = 0.09 \text{ lagging}$$

$$= 20 \angle -\cos^{-1}(0.09) \quad \text{... Neglecting sign as lagging}$$

$$= 20 \angle -84.83^\circ \text{ A } 1.8 - j19.9188A$$

$$\therefore \quad \bar{I}_1 = \bar{I}_0 + \bar{I}_{2r} = 1.8 - j19.9188A + 48.495 - j7.9873$$

$$= 50.295 - j27.9061 = 57518 \angle -29.02^\circ\ A$$

(*i*) Input current = 57.518 A

Input p.f. = cos (–29.02)° = 0.8744 lagging

(*ii*) $$T = \frac{\dfrac{3(I'_{2r})^2 R'_2}{S}}{\left(\dfrac{2\pi \times 750}{60}\right)}$$ Total torque developed.

$$T = \frac{\dfrac{3 \times (49.1484)^2 \times 0.15}{0.03}}{\left(\dfrac{2\pi \times 750}{60}\right)} = 461.3389 \text{ N-m.}$$

(*iii*) $$P_2 = \frac{3(I'_{2r})^2 R'_2}{5} = \frac{3 \times (49.1484)^2 \times 0.15}{0.03}$$

$P_2 = 36233.4783$ W Total as multiplied by 3

Now $P_2 : P_c : P_m$ is $1 : 5 : 1 - s$

$$\therefore \quad \frac{P_2}{P_m} = \frac{1}{1-s}$$

$$\therefore \quad P_m = (1 - s)\, P_2 = (1 - 0.003) \times 36233.4783$$

$$P_m = 35146.474 \text{ W.}$$

$$\therefore \quad \text{Total } P_{out} = P_m - \text{Mechanical loss}$$

$$P_{out} = 35146.474 - 1000$$

$$P_{out} = 34146.474 \text{ W.}$$

(*iv*) For efficiency, find the total losses,

Core loss = 1250 W

$$\text{Total copper loss} = 3 \times (I'_{2r})^2 R_2 = \text{stator} + \text{Rotor}$$
$$= 3 \times (49.1484)^2 \times 0.25$$
$$= 1811.674 \text{ W.}$$

$$\text{Mechanical loss} = 1000 \text{ W.}$$

$$\% \eta = \frac{P_{out}}{P_{in}} \times 100 = \frac{P_{out}}{P_{out} + \text{total losses}} \times 100$$

$$\% \eta = \frac{34146.474}{34146.474 + 1250 + 1811.674 + 1000} \times 100$$

$$\% \eta = 89.36\%$$

Example 1.17

A 3 phase, 200 V, induction motor has stator impedance of (0.07 + j0.4) Ω/ph while equivalent rotor impedance referred to stator is (0.08 + j0.2) Ω/ph Neglecting no load current, calculate the maximum mechanical power output and the slip corresponding to maximum output condition. Assume star connected stator.

Solution: The given values are

$$R_1 = 0.7\ \Omega, \quad X_1 = 0.4\ \Omega\ R'_2 = 0.08\ \Omega, \quad X'_2 = 0.2\ \Omega$$
$$R_{1e} = R_1 + R'_2 = 0.07 + 0.08 = 0.15\ \Omega$$
$$X_{1e} = X_1 + X'_2 = 0.4 + 0.2 = 0.6\ \Omega$$
$$Z_{1e} = R_{1e} + X_{1e} = 0.15 + j0.6\ \Omega$$
$$|Z_{1e}| = \sqrt{R_{1e}^2 + X_{1e}^2} = \sqrt{(0.15)^2 + (0.6)^2} = 0.6184\ \Omega$$
$$V_1/ph = \frac{(V_1)\text{ line}}{\sqrt{3}} = \frac{200}{\sqrt{3}} = 115.47 \text{ V.}$$

$$\therefore \quad (P_{out})_{max} = \frac{3V_1^2}{2(R_{1e} + Z_{1e})} = \frac{3 \times (115.47)^2}{2[0.15 + 0.6184]}$$

The corresponding slip is given by $s = \dfrac{R'_2}{R'_2 + Z_{1e}} = \dfrac{0.08}{0.08 + 0.6184}$

$$s = 0.1145$$

So slip is 11.45%.

Example 1.18

A 4 pole, 30 H.P., three phase, 400 V. 50 Hz induction motor operates at an efficiency of 0.8 with a p.f. of 0.75 lagging. Calculate the current by the motor from the mains.

Solution: The given values are

$$P = 4, f = 50 \text{ Hz},$$
$$P_{out} = 30 \text{ H.P.}, V_L = \text{Line voltage} = 400 \text{ V.}$$
$$\cos\phi = 0.75, \eta = 0.8$$
$$P_{out} = 30 \times 735.5 = 22065 \text{ W.}$$

$$\eta = \frac{P_{out}}{P_{in}} = \frac{22065}{0.8} = 27581.25W.$$

Now $$P_{in} = \sqrt{3}\, V_L I_L \cos\phi$$

$$27581.25 = \sqrt{3} \times 400 \times I_L \times 0.75$$

$\therefore$ $$I_L = 53.08$$ Current drawn from mains.

Example 1.19

A 4-pole, 3-phase induction motor delivers 37 H.P. at the shaft at a speed of 1425 r.p.m. on 500V, 50Hz supply. The mechanical losses total 3H.P. and the p.f. is 0.9. Calculate for this load (i) slip (ii) rotor Cu loss (iii) total power input if the stator loss in 2500 W (iv) efficiency (v) number of complete cycles per min of the rotor e.m.f.

Solution: The given values are

$$P = 4,\ P_{out} = 37 \text{ H.P.} \quad N = 1425 \text{ r.p.m. } f = 50 \text{ Hz.}$$

$$\text{Mechanical loss} = 3 \text{ H.P., } V_L = 500 \text{ V, } \cos\phi = 0.9$$

$$\text{Stator loss} = 2500 \text{ W.}$$

(*i*) $$N_s = \frac{120f}{P} = \frac{120 \times 50}{4} = 1500 \text{ rpm}$$

$$\therefore \quad s = \frac{N_s - N}{N_s} = \frac{1500 - 1425}{1500} = 0.05$$

(*ii*) P_{out} = 37 H.P. = 37 × 737.5 = 27287.5 W.

$P_m = P_{out}$ + Mechanical loss = Gross mechanical power developed.

Mechanical loss = 3H.P. = 3 × 737.5 = 2212.5 W.

$\therefore$ P_m = 27287.5 + 2212.5 = 29500 W.

Now $P_2 : P_c : P_m$ is $1 : 5 : 1 - s$

$$\therefore \quad \frac{P_c}{P_m} = \frac{s}{1-s}$$

$$\therefore \quad P_c = \left(\frac{s}{1-s}\right) P_m = \frac{0.05}{(1-0.05)} \times 29500$$

P_c = 1552.6315 W = Rotor copper loss

(*iii*) $$\frac{P_2}{P_c} = \frac{1}{5}$$

$$\therefore \quad P_2 = \frac{P_c}{s} = \frac{1552.6315}{0.05} = 31052.6315 \text{ W}$$

$$\therefore \quad P_{in} = P_2 + \text{Stator loss} = 31052.6315 + 2500$$

$$P_{in} = 33552.6315 \text{ W}$$

$$P_{in} = \text{Total power input}$$

(iv) $\% \eta = \frac{P_{out}}{P_{in}} \times 100$

$$\% \eta = \frac{27287.5}{33552.6315} \times 100 = 81.32\%$$

(v) $f_r = s_f = 0.05 \times 50 = 2.5$ Hz = 2.5 cycles/sec

$\therefore f_r = 2.5 \times 60 = 150$ cycle per min.

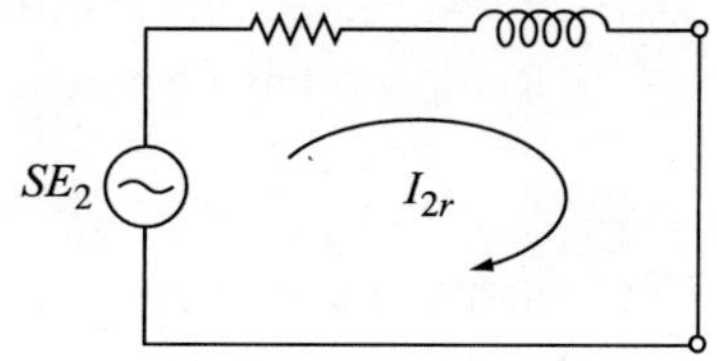

Fig. 1.27

Example 1.20

A 3-phase induction motor with star connected rotor has in induced electromotive force of 60 V between slip rings at standstill on open circuit with normal voltage applied to the stator. The resistance and standstill leakage reactance of each rotor phase are 0.6 Ω and 4 Ω respectively. Calculate the current per phase in the rotor.

(i) When it is at standstill and connected to a resistance of 5Ω and reactance of 2Ω per phase and

(ii) When running with a slip of 4% with its rotor terminals short circuited.

Solution: The given values are,

The emf between the slip rings = $E_{2\,line} = 60$ V.

$R_2 = 0.6\ \Omega, \quad X_2 = 4\Omega$

(i) Rotor stand still and $R_X = 5\ \Omega$, $X_X = 2\ \Omega$ added externally

$$E_{2\,ph} = \frac{E_{2\,line}}{\sqrt{3}}$$

$$= \frac{60}{\sqrt{3}} = 34.641 \text{ V.}$$

Total impedance on standstill is,

$$Z_T = R_2 + jX_2 + R_X + jX_X = 0.6 + j4 + 5 + j2$$

$$Z_T = 5.6 + j6\ \Omega$$

$$\therefore \quad I_2 = \frac{E_2}{Z_T} = \frac{34.641}{\sqrt{(5.6)^2 + (6)^2}} = 4.2207 \text{ A}$$

This is rotor current per phase.

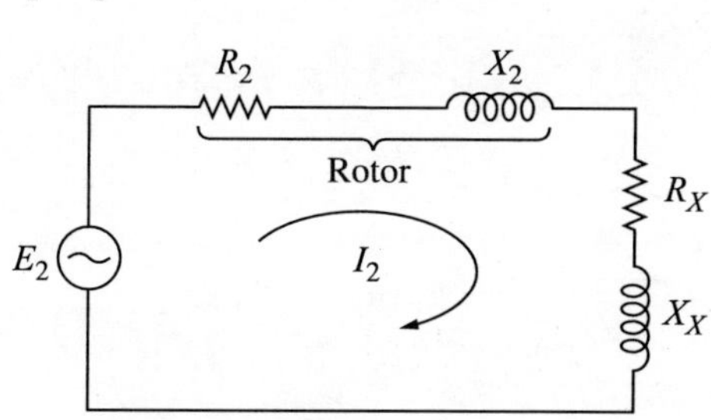

Fig. 1.28

(*ii*) $s = 4\%$ *i.e.*, 0.04 and rotor terminals shorted.

Rotor terminals shorted means external resistance and reactance completely removed.

$$Z_{2r} = R_2 + sX_2 \ \Omega$$

and e.m.f. $= sE_2$

$$\therefore \quad I_{2r} = \frac{E_{2r}}{Z_{2r}} = \frac{sE_2}{\sqrt{R_2^2 + (SX_2)^2}}$$

$$I_{2r} = \frac{0.04 \times 34.641}{\sqrt{(0.6)^2 + (0.04 \times 4)^2}} = 2.2314 \text{ A.}$$

Example 1.21

A 4 pole, 50 Hz, 3 phase induction motor, when loaded has total rotor input of 3000 W. Determine the torque developed in (i) N-m and (ii) syn.watt.

Solution: The given values are

$$P = 4, f = 50 \text{ Hz}, \quad P_2 = 3000 \text{ W}$$

$$N_s = \frac{120 f}{P} = \frac{120 \times 50}{4} = 1500 \text{ r.p.m}$$

$$T = \frac{P_2}{\frac{2\pi N_s}{60}} \text{ N-m} = \frac{3000}{\frac{2\pi \times 1500}{60}} = 19.098 \text{ N-m}$$

Now $$1 \text{ N-m} = \frac{2\pi Ns}{60} \text{ syn. watt}$$

$$\therefore \quad T \text{ in syn. watt} = 19.098 \times \frac{2\pi N_s}{60} = 3000 \text{ syn. watt.}$$

Example 1.22

A 1000 V, 50 Hz, 3 phase slip ring induction motor has star connected stator. The ratio of stator to rotor turn is 3.6. The rotor standstill impedance is 0.01 + j 0.2 Ω. Calculate the external resistance per phase required in rotor circuit to limit starting rotor current to 200 A.

Solution: The given values are

$$(E_1)_{\text{line}} = 1000 \text{ V}, R_2 = 0.01 \ \Omega, X_2 = 0.2 \text{ W}, (I_2)_{\text{st}} = 200 \text{ A}$$

$$K = \frac{(E_2)_{\text{ph}}}{(E_1)_{\text{ph}}} = \frac{\text{Rotor turns}}{\text{Stator turns}} = \frac{1}{3.6}$$

Now $$(E_1)_{\text{ph}} = \frac{(E_1)\text{line}}{\sqrt{3}} = \frac{1000}{\sqrt{3}} = 577.35 \text{ V}$$

$$\frac{E_2}{E_1} = \frac{1}{3.6} = \frac{E_2}{577.35} \quad \textit{i.e.,} \quad E_2 = 160.375 \text{ V}$$

Let as start R_X be the external resistance per phase, introduced in rotor circuit,

$$\therefore \quad R_2' = R_2 + RX$$

$$\therefore \quad (I_2)_{st} = \text{Rotor current at start} = \frac{E_2}{\sqrt{R_2'^2 + X_2^2}}$$

$$\therefore \quad 200 = \frac{160.375}{\sqrt{(R_2')^2 + (0.2)^2}}$$

$$\therefore \quad (R_2')^2 + (0.2)^2 = 0.6429 \qquad i.e., (R_2')^2 = 0.6029$$

$$\therefore \quad R_2' = 0.7765\ \Omega$$

$$R_X = 0.7765 - 0.01 = 0.76665\ \Omega \text{ per phase.}$$

Example 1.23

A three phase, 50 Hz, 400 V induction motor has 4 poles star connected stator winding. Rotor resistance and reactance per phase are 0.15 Ω and 1 Ω respectively. Full load slip is 5%. Calculate (a) Total torque developed (b) Maximum torque (c) Speed at maximum torque. Assume stator to rotor ratio 2 : 1.

Solution: The given values are

50 Hz, $V_L = 400$ V, $P = 4$, $R_2 = 0.15\ \Omega$, $X_2 = 1\ \Omega$, $s = 0.05$

Now, E_1 (line) = $V_L = 400$ V.

$$\therefore \quad E_{1\text{ph}} = \frac{E_1(\text{line})}{\sqrt{3}} = \frac{400}{\sqrt{3}} = 230.94 \text{ V}$$

$$\frac{E_2}{E_1} = \frac{1}{2}$$

$$\therefore \quad E_2 = \frac{E_{1\text{ph}}}{2} = \frac{230.94}{2} = 115.47 \text{ V}$$

$$N_s = \frac{120f}{P} = \frac{120 \times 50}{4} = 1500 \text{ r.p.m}$$

$$n_s = \frac{N_s}{60} = 25 \text{ r.p.m}$$

$$(i)\ T = \frac{3}{2\pi n_s} \times \frac{sE_2^2 R_2}{R_2^2 + (sX_2)^2} = \frac{3}{2\pi \times 25} \times \frac{0.05 \times (115.47)^2 \times 0.15}{[(0.15)^2 + (0.05 \times 1)^2]} = 76.39 \text{ N-m}$$

$$(ii)\ T_m = \frac{3}{2\pi n_s} \times \frac{E_2^2}{2X_2} = \frac{3}{2\pi \times 25} \times \frac{(115.47)^2}{2 \times 1} = 127.323 \text{ N-m.}$$

$$(iii)\ S_m = \frac{R_2}{X_2} = \frac{0.15}{1} = 0.15$$

$$\therefore \quad N = (1 - S_m)\, N_s = (1 - 0.15) \times 1500 = 1275 \text{ r.p.m.}$$

Example 1.24

A 400 V, 50 Hz, 3 phase, 6 pole star connected induction motor has the following values for the various parameters of its equivalent circuit.

Stator impedance = (0.3 + j0.4) Ω

Equivalent rotor impedance = (0.3 + j0.4) Ω

Equivalent rotor impedance = (0.2 + j0.4) Ω

Magnetising reactance = 20 Ω

Resistance to account fore core loss = 100 Ω.

Using the approximate equivalent circuit calculate for a slip of 4% (a) motor speed (b) stator current (c) power factor (d) motor output and efficiency.

Assume stator losses, mechanical losses to 2 kW each.

Solution: The given values are

$R_1 = 0.3\ \Omega$ $\quad X_1 = 0.4\ \Omega$ Mechanical losses = 2 kW

$R_2' = 0.2\ \Omega$ $\quad X_1' = 0.4\ \Omega$

$X_0 = 20\ \Omega$ $\quad S = 0.04$

$R_0 = 100\ \Omega$ $\quad$ Stator loss = 2 kW

The approximate equivalent circuit is shown in Fig. 1.29.

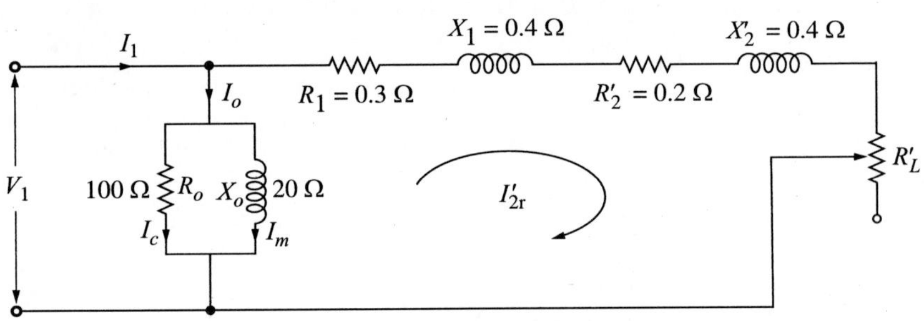

Fig. 1.29

$$V_1 = \frac{400}{\sqrt{3}} = 230.94 \text{ volts}, \quad N_s = \frac{120f}{P} = \frac{120 \times 50}{6} = 1000 \text{ rpm}$$

$$V_1 = \frac{R_2'(1-s)}{s} = \frac{0.2(1-0.04)}{(0.04)} = 4.8\ \Omega$$

$$R_{1e} = R_1 + R_2' = 0.3 + 0.2 = 0.5\ \Omega$$

$$X_{1e} = X_1 + X_2' = 0.4 + 0.4 = 0.8\ \Omega$$

$$I_{2r}' = \frac{V_1}{(R_{1e} + Ri) + jX_{1e}} = \frac{230.94}{(0.5 + 4.8) + j0.8}$$

$$I_{2r}' = \frac{230.94}{5.3 + j0.8} = \frac{230.94}{5.36\ \angle 8.58^\circ}$$

$$I_{2r}' = 43.08\ \angle -8.58^\circ \text{ A} = 42.59 - j6.42 \text{ A}$$

$$I_c = \frac{V_1}{R_0} = \frac{230.94}{100} = 2.3094 \text{ A}$$

$$I_m = \frac{V_1}{jX_0} = \frac{230.94}{jZ_0} = 11.547 \text{ A } \angle{-90°}\ A = -\,j11.547$$

$$\therefore \quad I_0 = (2.3094 - j11.547)\ A = 11.77\ \angle{-78.69}\ A$$

$$\text{Core loss/ph} = I_c^2 R_0 = (2.3094)^2\ (100) = 533.33 \text{ W.}$$

$$\text{Stator current, } \overline{I}_1 = \overline{I}_0 + \overline{I}_{2r}$$

$$= (2.3094 - j11.547) + (42.59 - \text{j}6.42)$$

$$= (44.8994 - \text{j}17.967) \text{ A}$$

$$= 48.36\ \angle{-21.80}\ A$$

$$\text{Stator current} = 48.36 \text{ A}$$

$$\text{P.f.} = \cos \phi_1 = \cos (21.80) = 0.9284 \text{ (lagging)}$$

$$P_c = \text{Rotor copper less} = 3\ (I'_{2r})^2\ R'_2 = 3\ (42.08)^2\ (0.2)$$

$$= 1113.53 \text{ W.}$$

Now $P_2 : P_0 : P_m$ is $1 : s : 1 - s$

$$\frac{P_c}{P_m} = \frac{s}{1-s}$$

$$\therefore \quad P_m = \left(\frac{1-s}{s}\right) P_c = \frac{1-0.04}{0.04}(1113.53)$$

$$P_m = 26724.72 \text{ W}$$

$$P_{out} = P_m - \text{Mechanical losses}$$

$$P_{out} = 26724.72 - 2000$$

$$P_{out} = 24724.72 \text{ W}$$

$$\text{Stator copper losses} = 3I_1^2 R_1$$

$$= 3(48.36)^2\ (0.3)$$

$$= 2104.82 \text{ W.}$$

$$\text{Total loss} = \text{core less} + \text{stator } Cu \text{ loss} + Pc + \text{Mechanical loss}$$

$$= (3 \times 533.33) + (2104.82) + (1113.53) + (2000)$$

$$= 6818.34 \text{ W.}$$

$$\%\ \text{Efficiency} = \frac{P_{out}}{P_{in}} \times 100 = \frac{P_{out}}{P_{out} + \text{Losses}} \times 100$$

$$= \frac{24724.72}{24724.72 + 6818.34} \times 100 = 78.38\%$$

$$\text{Motor speed, } N = N_s(1 - s) = 1000\ (1 - 0.04) = 1000\ (0.96)$$

$$N = 960 \text{ r.p.m}$$

Example 1.25

Compute the value of maximum torque in N-m for a 3-phase, 440 V, star connected, 50 Hz, 4 pole induction motor for which $X_1 = 5.2\ \Omega$, $R_2' = 1.2\ \Omega$ and $X_2' = 4.5\ \Omega$. Also determine the slip corresponding to it.

Solution: $R_2' = 1.2\ \Omega$, $X_2' = 4.5\ \Omega$, $X_1 = 5.2\ \Omega$

The approximate equivalent circuit is shown in Fig. 1.30.

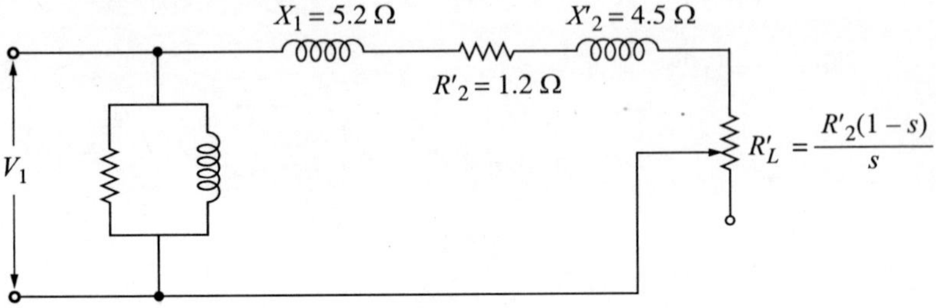

Fig. 1.30

The mechanical torque developed by rotor would be maximum when there is maximum transfer of power to the resistor R_1'. The above circuit can also be redrawn as shown in the Fig. 1.31.

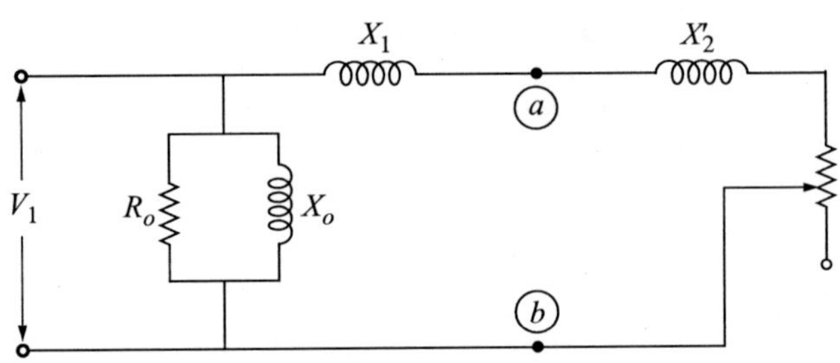

Fig. 1.31

The we find the Thevenin's circuit across points (a) and (b), we get

The condition for maximum power after matching the impedance is given by

$$\frac{R_2'}{S} = \sqrt{R_{TH}^2 + (X_{TH} + X_2')^2}$$

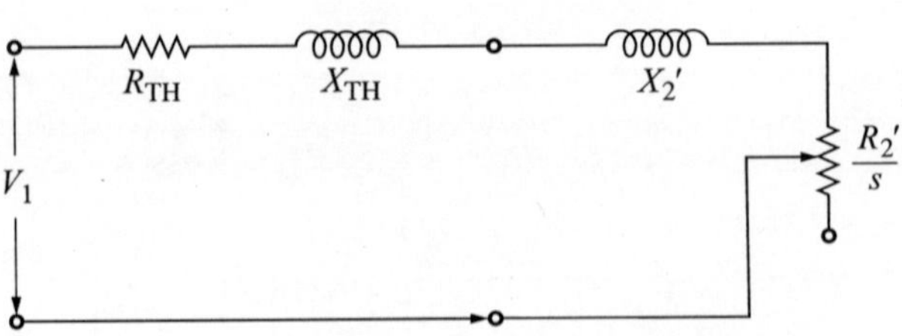

Fig. 1.32

Here magnetising branch components are not given hence,

$$R_{TH} = R_1,\ X_{TH} = X_1$$

But R_1 is also not given

$$\therefore \qquad \frac{R_2'}{S_m} = \sqrt{(X_1 + X_2')^2} = \sqrt{(5.2+4.5)^2} = 9.7$$

$$\therefore \qquad S_m = \frac{R_2'}{9.7} = 0.1237 \quad \text{or} \quad 12.37\%$$

The maximum torque can be obtained as follows:

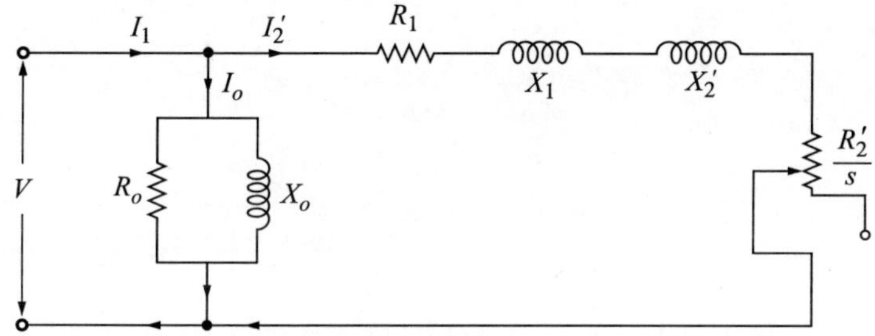

Fig. 1.33

Neglecting I_0 we have

$$I_1 = I_2'$$

$$\therefore \qquad I_2' = \frac{V}{\sqrt{\left(R_2 + \frac{R_2'}{s}\right)^2 + (X_1 + X_2')^2}} = \frac{V}{\sqrt{\left(\frac{R_2'}{S_m}\right)^2 + (X_1 + X_2')^2}}$$

$$= \frac{(4401\sqrt{3})}{\sqrt{(9.7)^2 + (5.2+4.5)^2}} = \frac{254.03}{13.71} = 18.51 \text{ A}$$

$$N_s = \frac{120f}{P} = \frac{120 \times 50}{4} = 1500 \text{ r.p.m}$$

$$T_{max} = \frac{3I_2'R_2'/S_m}{2\pi N_s/60} = \frac{3(18.51)^2(9.7)}{\left[\frac{2\pi \times 1500}{60}\right]}$$

$$\therefore \qquad T_{max} = 63.47 \text{ N-m.}$$

Example 1.26

A four pole, 50 Hz three phase wound rotor induction motor has rotor resistance per phase of 0.02 ohm and 0.1 ohms respectively. What must be the value of the external resistance per phase to be inserted in the rotor circuit such that the starting torque is two-thirds of the maximum torque.

Solution: $P = 4, f = 50$ Hz, $R_2 = 0.02\ \Omega$, $X_2 = 0.1\Omega$, $T_{st} = \dfrac{2}{3}T_m$

$$T_{st} \propto \frac{E_2^2 R_2'}{R_2'^2 + X_2^2} \qquad s = 1 \text{ at start}$$

$$T_m \propto \frac{E_2^2}{2 \times 2}$$

$$\therefore \quad \frac{T_{st}}{T_m} = \frac{E_2^2 R_2'}{R_2'^2 + X_2^2} \times \frac{2X_2}{E_2} = \frac{2R_2' \times 2}{R_2'^2 + X_2^2}$$

$$\therefore \quad \frac{2}{3} = \frac{2R_2' \times 0.1}{(R_2')^2 + (0.1)^2}$$

$$\therefore \quad (R_2')^2 - 0.3\ R_2' + 0.01 = 0$$

$$R_2' = \frac{0.3 \pm \sqrt{(0.3)^2 - (4 \times 0.01)}}{2} = 0.2418, 0.038$$

Neglecting higher value as R_2' cannot be greater than X_2.

$$R_2' = 0.038\ \Omega = R_2 + R_{ex}$$

$$\therefore \quad R_{ex} = 0.038 - R_2 = 0.038 - 0.02 = 0.018\ \Omega$$

This is the extra resistance required.

Example 1.27

A 3300 V, 10 pole, 50 Hz, 3-phase star connected induction motor has slip rotor resistance per phase = 0.015 ohm and standstill reactance per phase = 0.25 Ω if the motor runs at 2.5% slip on full load, find.

1. *The speed of the motor*
2. *Speed of the motor*
3. *The ratio of maximum torque to full load torque.*

Solution: $V_L = 3300$ V, $P = 10, f = 50$ Hz, $R_2 = 0.015\ \Omega$, $X_2 = 0.25\ \Omega$, $s_{fl} = 2.5\%$

1. $N_s = \dfrac{120f}{P} = \dfrac{120 \times 50}{10} = 600$ r.p.m.

   $\therefore \quad N = N_s(1 - s_{fl}) = 600\ (1 - 0.025) = 264$ r.p.m.

2. $S_m = \dfrac{R_2}{X_2} = \dfrac{0.015}{0.25} = 0.06$ *i.e.,* 6%

   $\therefore \quad N_m = N_s(1 - s_m) = 600\ (1 - 0.06) = 564$ r.p.m

3. $T_{fl} \propto \dfrac{S_{f.L.} R_2}{R_2^2 + (S_{fl} X_2)^2} \times \dfrac{R_2^2 + (S_{fl} X_2)^2}{S_{fl} R_2}$

$$\therefore \quad \frac{T_m}{T_{fl}} = \frac{s_m R_2}{R_2^2 + (s_m X_2)^2} \times \frac{R_2^2 + (s_{fl} X_2)^2}{s_{fl} R_2}$$

$$= \frac{0.06 \times [(0.015)^2 + (0.025 \times 0.25)^2]}{0.025 \times [(0.015)^2 + (0.06 \times 0.25)^2]}$$

$$\therefore \quad \frac{T_m}{T_{fl}} = 1.4083$$

Example 1.28

The following data refers to a 10-pole, 400V, 50 Hz, 3-phase induction motor.

$R_1 = 1.75\ \Omega$, $X_1 = 5.5\ \Omega$, $R_2' = 2.25\ \Omega$, $X_2' = 6.6\ \Omega$. When the motor is tested on no load, it is observed that it takes 3.8 A (line current) and the total core loss is 310 W. By using an approximate equivalent circuit at 4% slip, calculate:

- *the rotor current*
- *supply current and power factor*
- *mechanical power developed*
- *gross load torque*

Also determine the equivalent circuit.

Solution: $I_0 = 3.8\ A$, $S = 4\%$, $W_0 = 310$ W, $P = 10$, $V_L = 400$ V

Assuming star connection

$$W_0 = \sqrt{3}\, V_0 I_0 \cos\phi_0$$

$$\therefore \quad 310 = \sqrt{3} \times 400 \times 3.8 \times \cos\phi_0$$

$$\therefore \quad \cos\phi_0 = 0.1177, \quad \sin\phi_0 = 0.993$$

$$\therefore \quad I_c = I_0 \cos\phi_0 = 0.4474 \text{ A}$$

and $$I_m = I_0 \sin\phi_0 = 3.7734 \text{ A}$$

$$\therefore \quad R_0 = \frac{V_{ph}}{I_m} = \frac{400/\sqrt{3}}{0.4474} = 516.1826\ \Omega$$

and $$X_0 = \frac{V_{ph}}{I_m} = \frac{400/\sqrt{3}}{3.7734} = 61.202\ \Omega$$

$$V_1 = \frac{400}{\sqrt{3}} = 230.94 \text{ V} \qquad \text{...Assuming star connection}$$

$$N_s = \frac{120 f}{P} = \frac{120 \times 50}{10} = 600 \text{ r.p.m}$$

$$R_L' = \frac{R_L'(1-s)}{s} = \frac{2.25(1-0.04)}{0.04} = 54\ \Omega$$

$$R_{1e} = R_1 + R_2' = 1.75 + 2.25 = 4\ \Omega$$

$$X_{1e} = X_1 + X_2' = 5.5 + 6.6 = 12.1\ \Omega$$

$$I_{2r}' = \frac{V_1}{(R_{1e} + R_L') + jX_{1e}}$$

$$= \frac{230.94}{(4+54)+j12.1} = \frac{230.94}{58.248+ \angle 11.78°}$$

Rotor current, $I'_{2r} = 3.897 \angle -11.78° A$

$$I_0 = I_C - jI_m = 0.4474 - j3.7734 \text{ A}$$

$\therefore$ $$\bar{I}_1 = \bar{I}_0 + \bar{I}'_{2r} = (0.4474 - j3.7734) + (3.897 \angle -11.78°)$$

$$= (0.4474 - j_3.7734) + (3.8149 - j_0.796)$$

Supply current $I_1 = 4.2623 - j4.5694 \text{ A} = 6.24 \angle -46.99° A$

$\therefore$ $\cos\phi = \cos(-46.99°) = 0.6821$ lagging ... Power factor

$$P_c = 3(I'_{2r})2\ R'^2 = 3 \times (3.897)2 \times 2.25 = 102.5096 \text{ W}$$

Now $P_2 : P_c : P_m$ is $1 : s : 1 - s$

$\therefore$ $$\frac{P_c}{P_m} = \frac{s}{1-s}$$

$\therefore$ $$P_m = \frac{102.5096 \times (1-0.04)}{0.04} = 2460.23 \text{ W}$$

...Mechanical power

$$N = N_s(1 - s) = 600 \times (1 - 0.04) = 576 \text{ r.p.m}$$

$\therefore$ $$T = \frac{Pm}{w} = \frac{2460.23}{\left(\frac{2\pi N}{60}\right)} = 40.787 \text{ N-m}$$

The equivalent circuit is as shown.

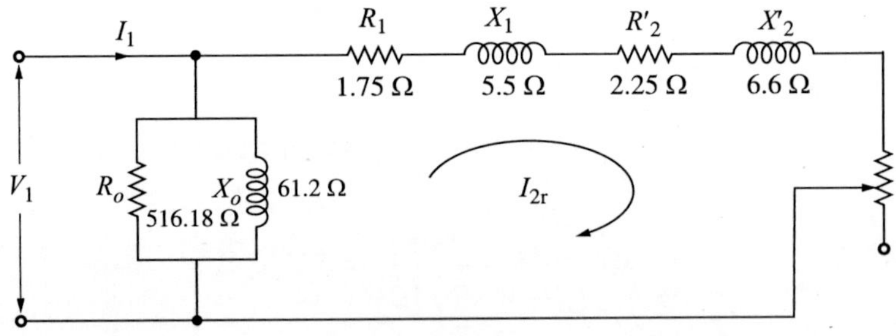

Fig. 1.34

Example 1.29

A 12 pole, 3 phase alternator is coupled to an engine running at 500 r.p.m. of supplies an induction motor which has a full-load speed of 1440 r.p.m. Find the slip of the motor.

Solution:

$$P_A = 12$$

$$N_s = 500 \text{ r.p.m}$$

$$N_s = 120 f/PA \qquad \text{for alternator}$$

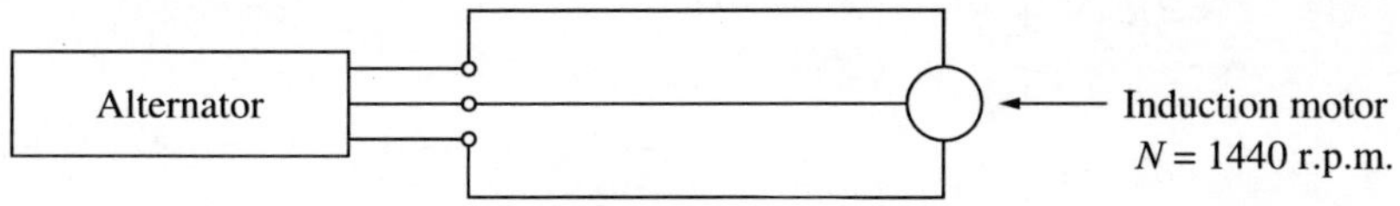

Fig. 1.35

$$\therefore \quad 500 = \frac{120f}{12}$$

$$\therefore \quad f = 50 \text{ Hz.}$$

For motor $\quad N_s = \dfrac{120f}{P_M} \quad$ ...Assume $P_M = 4$ for motor

$$\therefore \quad N_s = \frac{120 \times 50}{4} = 1500 \text{ r.p.m}$$

$$\therefore \quad \% \, s = \frac{N_s - N}{N_s} \times 100 = \frac{1500 - 1440}{1500} \times 100 = 4\%$$

Example 1.30

Rotor resistance and standstill reactance per phase of a phase induction motor are 0.04 Ω and 0.2 Ω respectively. What should be external respectively. What should be external resistance required at start in rotor circuit to obtain

(i) maximum torque at start

(ii) 50% of maximum torque at start.

Solution: The given values all

$$R_2 = 0.04 \, \Omega \quad \text{and} \quad X_2 = 0.2 \, \Omega$$

(*i*) To get maximum torque at start, $s_m = 1$.

$$\therefore \quad s_m = \frac{R_2}{X_2} = 1.$$

$$\therefore \quad R_2 = X_2 = 0.2 \, \Omega$$

$$\therefore \quad R_X = 0.2 - 0.04 = 0.16 \, \Omega$$

This is the external resistance required.

(*ii*) $T_{st} = 0.5 \, T_m$

Now $T_{st} \propto \dfrac{E_2^2 R_2'}{R_2'^2 + X_2^2}$ and $T_m \propto \dfrac{E_2^2 R_2'}{R_2'^2 + X_2^2}$

where R_2' is new rotor resistance due to R_X.

$$\therefore \quad \frac{T_{st}}{T_m} = \frac{E_2^2 R_2'}{R_2'^2 + X_2^2} \quad \text{and} \quad T_m \propto \frac{E_2^2}{2X_2}$$

$$\therefore \quad 0.5 = \frac{2R_2'(0.2)}{(R_2')^2 + (0.2)^2}$$

$\therefore \quad (R'_2)^2 + (0.04) = 0.8\, R'_2$

$\therefore \quad (R'_2)^2 - 0.8\, R'_2 + 0.04 = 0$

$$\therefore \quad R'_2 = \frac{0.8 \pm \sqrt{(0.8)^2 - 4 \times 0.04}}{2} = 0.0535\ \Omega,\ 0.7464\ \Omega$$

Now R'_2 cannot be greater than X_2

$\therefore \quad R'_2 = 0.0535 = R_2 + R_X$

$\therefore \quad R_X = 0.0535 - 0.4 = 0.0135\ \Omega$ This is the external resistance required.

Example 1.31

A 3 phase, 50 Hz, 8 pole induction motor has a full load slip of 40%. The rotor resistance is 0.001 Ω per phase and the standstill reactance 0.005 Ω per phase. Find the ratio of maximum to full load torque and the speed at which maximum torque occurs.

Solution: The given values are

$P = 8 \quad f = 50$ Hz, $\quad s_f = 40\% = 0.4$

$R_2 = 0.001$ Ω/ph

$X_2 = 0.005$ Ω/ph

Now $$T \propto \frac{sR_2}{R_2^2 + (sX_2)^2} \quad \text{with } E_2 \text{ constant}$$

$$T_{F.L.} \propto \frac{s_f R_2}{R_2^2 + (s_f X_2)^2} \quad \text{where } s_f \text{ is full load slip}$$

and $$T_m \propto \frac{s_m R_2}{R_2^2 + (s_m X_2)^2}$$

$$s_m = \frac{R_2}{X_2} = \frac{0.001}{0.005} = 0.2$$

Taking ratio of T_m and $T_{F.L.}$

$$\frac{T_m}{T_{F.L.}} = \frac{s_m R_2}{R_2^2 + (s_m X_2)^2} \cdot \frac{R_2^2 + (s_f X_2)^2}{s_f \cdot R_2}$$

$$= \frac{s_m \left[R_2^2 + (s_f X_2)^2 \right]}{s_f \left[R_2^2 + (s_m X_2)^2 \right]}$$

$$= \frac{0.2[(0.001)^2 + (0.4 \times 0.005)^2]}{0.4[(0.001)^2 + (0.2 \times 0.005)^2]}$$

$$\therefore \quad \frac{T_m}{T_{F.L.}} = 1.25$$

$s_m = 0.2$ = slip at which T_m occurs.

$$N_s = \frac{120\,f}{P} = \frac{120 \times 50}{8} = 750 \text{ rpm}$$

$$N = N_s(1 - s_m) = 750\,(1 - 0.2)$$

$$N = 600 \text{ rpm}$$

This is the speed at which maximum torque occurs.

Example 1.32

A 4 pole, 400 V, 3 phase induction motor has a standstill rotor emf of 100 V per phase. The rotor has a resistance of 50 MΩ per phase. Calculate the maximum torque and the slip at which it occurs. Neglect stator impedance.

Solution: The given values are

$P = 4$, $V_L = 400$ V, $E_2 = 100$ V, $R_2 = 50$ mΩ $= 0.05\ \Omega$ $X_2 = 0.5\ \Omega$

The torque occurs when

$$s_m = \frac{R_2}{X_2} = \frac{0.05}{0.5} = 0.1$$

we have
$$T_{\max} = \frac{3}{2\pi n_s} \cdot \frac{s_m E_2^2 R_2}{R_2^2 + (s_m X_2)^2}$$

$$= \frac{3}{2\pi n_s} \cdot \frac{(R_2/X_2)\,E_2^2 \cdot R_2}{R_2^2 + (R_2/X_2 . X_2)^2}$$

$$= \frac{3}{2\pi n_s} \frac{1}{X_2} \cdot \frac{R_2^2 E_2^2}{2R_2^2} = \frac{3}{2\pi n_s} \cdot \frac{E_2^2}{2X_2}$$

Now
$$n_s = \frac{120 f}{P} = \frac{120 \times 50}{4} = 1500 \text{ rpm} \quad \text{i.e. } n_s = \frac{1500}{60} = 25 \text{ rps}$$

$$T_{\max} = \frac{3}{2\pi \times 25} \cdot \frac{(100)^2}{2(0.5)} = \frac{(3)\,(100)^2}{50 \times \pi}$$

$$T_{\max} = 190.98 \text{ N-m}$$

Example 1.33

A 3-phase induction motor has a starting torque of 100% and maximum torque of 200% of the full load torque. Determine,

(i) Slip at maximum torque occurs

(ii) Full load slip.

Solution: T_{st} = 100% of T_{fl} and T_m = 200% of T_{fl}.

$$\therefore \qquad \frac{T_{st}}{T_{fl}} = 1 \quad \text{and} \quad \frac{T_m}{T_{fl}} = 2$$

Now $T \propto \dfrac{sE_2^2 R_2}{R_2^2 + (sX_2)^2}$ and $T_{st} \propto \dfrac{E_2^2 R_2}{R_2^2 + X_2^2}$

(i) $\dfrac{T_{st}}{T_m} = \dfrac{2s_m}{s_m^2 + 1}$ and $\dfrac{T_{st}}{T_m} = \dfrac{1}{2}$ from given ratios.

$$\therefore \quad \frac{2s_m}{s_m^2 + 1} = \frac{1}{2} \quad i.e., \quad s_m^2 - 4s_m + 1 = 0 \quad i.e., \; s_m = 3.732,\ 0.2679$$

$$\therefore \quad s_m = 0.2679 \quad i.e., \quad 26.79\%$$

(ii) Using equation (2) and $s_m = 33.33\%$

$$2 = \frac{s_m\left[R_2^2 + (s_{fl} X_2)^2\right]}{s_{fl}\left[R_2^2 + (s_m X_2)^2\right]} = \frac{\left[\left(\dfrac{R_2}{X_2}\right)^2 + (s_{fl})^2\right] s_m}{\left[\left(\dfrac{R_2}{X_2}\right)^2 + (s_m)^2\right] s_{fl}}$$

$$\therefore \quad 2 = \frac{s_m^2 + s_{fl}^2}{2 s_m s_{fl}}$$

$$\therefore \quad 2 = \frac{(0.2679)^2 + (s_{fl})^2}{2 \times (0.2679) \times s_{fl}}$$

$$\therefore \quad s_{fl}^2 - 1.0716\, s_{fl} + 0.07177 = 0$$

$$\therefore \quad s_{fl} = 0.07178 \quad i.e., \quad 7.178\%$$

Example 1.34

A 8 pole, 50 Hz, 3-phase, induction motor develops a maximum torque of 150 N-m at 650 rpm. The rotor resistance is 0.6 ohm/phase. Find torque at 4% slip neglect stator impedance.

Solution: $P = 8$, $f = 50$ Hz, $T_m = 150\ N_m$ at $N = 650$ rpm

$R_2 = 0.6\ \Omega$, $s = 4\% = 0.04$

$$N_s = \frac{120 f}{P} = \frac{120 \times 50}{8} = 750 \text{ rpm}$$

$$\therefore \quad s_m = \frac{N_s - N}{N_s} = \frac{750 - 650}{750} = 0.1333$$

Now $\quad T_m \propto \dfrac{s_m E_2^2 R_2}{R_2^2 + (s_m X_2)^2}$ and $T \propto \dfrac{sE_2^2 R_2}{R_2^2 + (sX_2)^2}$

and $\quad s_m = \dfrac{R_2}{X_2} = 0.1333$

$$\therefore \quad X_2 = \frac{0.6}{0.1333} = 4.5011\ \Omega$$

$$\therefore \quad \frac{T_m}{T} = \frac{\dfrac{s_m E_2^2 R_2}{R_2^2 + (s_m X_2)^2}}{\dfrac{s E_2^2 R_2}{R_2^2 + (s X_2)^2}} = \frac{s_m\left[R_2^2 + (sX_2)^2\right]}{s\left[R_2^2 + (s_m X_2)^2\right]}$$

$$\therefore \quad \frac{150}{T} = \frac{0.1333\,[(0.6)^2 + (0.04 \times 4.5011)^2]}{0.04[(0.6)^2 + (0.1333 \times 4.5011)^2]}$$

$$\therefore \quad \frac{150}{T} = 1.8163$$

$$\therefore \quad T = 82.5854 \text{ N-m} \qquad \text{torque at 4\% slip.}$$

Example 1.35

A three phase, star connected 400 volts, 50 Hz, 4 pole induction motor has following parameter referred to stator,

$r_1 = 0.15$ ohm, $X_2 = 0.44$ ohm

$r_2' = 0.12$ ohm $X_2' = 0.44$ ohm

$x_m = 30$ ohm neglect core loss resistance (r_c), find stator current, power factor at rated voltage and slip of 4%.

Solution: $r_1 = 0.15\ \Omega$, $X_1 = 0.44\ \Omega$, $r_2' = 0.12\ \Omega$, $X_2' = 0.44\ \Omega$

$X_m = 30\ \Omega$, $s = 0.04$

$$R_L' = \frac{r_2'(1-s)}{s} = \frac{0.12(1-0.04)}{0.04} = 2.88\ \Omega$$

The equivalent circuit is shown in Fig. 1.36.

$$V_1 = \frac{400}{\sqrt{3}} = 230.94 \text{ V}$$

$$I_{2r}' = \frac{V_1}{(r_1 + r_2' + R_L') + j(X_1 + X_2')} = \frac{230.94\angle 0^\circ}{3.15 + j0.88}$$

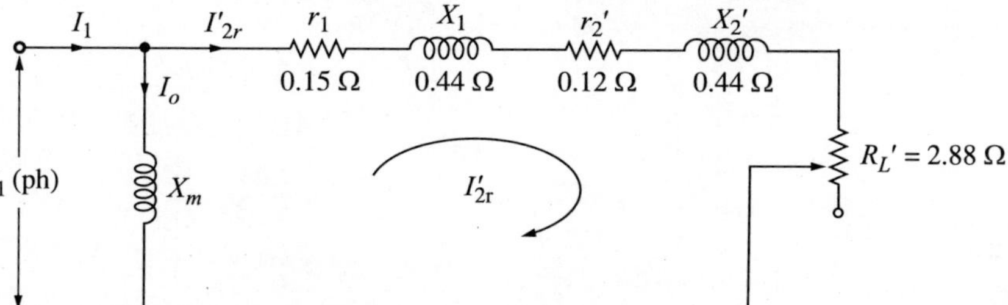

Fig. 1.36

$$= \frac{230.94\angle 0^\circ}{3.2706\angle 15.6^\circ} = 70.6108\angle -15.6^\circ \text{ A}$$

$$I'_{2r} = 68 - j\,18.98 \text{ A}$$

$$I_0 = \frac{V_1}{0 + jX_m} = \frac{230.94\angle 0^\circ}{0 + j30} = -\,j7.698 \text{ A}$$

$\therefore$ $$I_1 = I_0 + I'_{2r} = 68 - j\,26.678 \text{ A}$$

$\therefore$ $$I_1 = 73.045\angle -21.42^\circ \text{ A}$$ Stator current

$\therefore$ $$\cos\phi = \cos(-21.42^\circ) = 0.9309 \text{ lagging power factor}$$

Example 1.36

A 440 V, 50 Hz, 4-pole, 3-phase delta connected induction motor has a leakage impedance of (0.3 + j 5.5 + 0.25/s) ohm/phase (delta phase) referred to the stator. The stator to rotor voltage ratio is 2.5. Determine the external resistance to be inserted in each star-phase of the rotor winding such that the motor develops a gross torque of 150 N-m at a speed of 1250 rpm.

Solution: $V_L = 440$ V, $f = 50$ Hz, $P = 4$, delta connected

$$k = \frac{N_1}{N_2} = 2.5, \quad N = 1250 \text{ rpm}$$

$$N_s = \frac{120f}{P} = \frac{120 \times 50}{4} = 1500 \text{ rpm}$$

$$s = \frac{N_s - N}{N_s} = \frac{1500 - 1250}{1500} = 0.167$$

Total leakage impedance referred to stator (equivalent star bias)

$$= \frac{1}{3}\left[0.3 + \frac{0.25}{s} + j5.5\right]$$

$$= \left(0.1 + \frac{0.083}{s} + j1.83\right) \Omega\text{/ph (star)}$$

Consider the equivalent circuit shown in Fig. 1.37

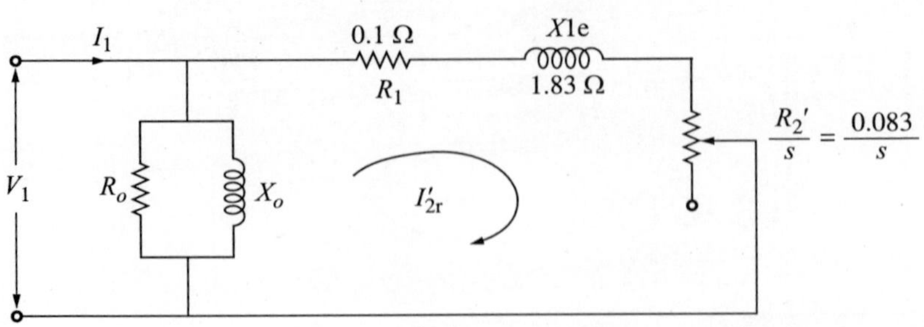

Fig. 1.37

Let R_x be the additional resistance inserted in each rotor phase when $s = 0.167$ and hence new rotor resistance be R_{2x}.

$$\therefore \quad I'_{2r} = \frac{V_1}{\left[0.1 + j1.83 + \dfrac{R_x}{s}\right]} = \frac{(440/\sqrt{3})}{\sqrt{\left(0.1 + \dfrac{R_{2x}}{0.167}\right)^2 + (1.83)^2}} \text{ A}$$

$$\therefore \quad T = \frac{1}{2\pi n_s} \times \frac{3(I'_{2r})^2 R_{2x}}{s}$$

$$\therefore \quad 150 = \frac{1}{2\pi\left(\dfrac{1500}{60}\right)} \times \frac{3\left(\dfrac{440}{\sqrt{3}}\right)^2 \times R_{2x}}{\left[\left(0.1 + \dfrac{R_{2x}}{0.167}\right)^2 + (1.83)^2\right]} \times \frac{1}{0.167}$$

$$\left(0.1 \times \frac{R_{2x}}{0.167}\right)^2 + (1.83)^2 = 49.2014\ R_{2x}$$

$$\therefore \quad 35.856\ R_{2x}^2 + 1.1976\ R_{2x} + 0.01 + 0.01 + 3.34.89 - 49.2014\ R_{2x} = 0$$

$$35.856\ R_{2x}^2 - 48\ R_{2x} + 3.3589 = 0$$

$$\therefore \quad R_{2x} = 1.2646,\ 0.074\ \Omega \text{ Neglecting lower value.}$$

$$R_x = R_{2x} - 0.083 = 1.2646 - 0.083 = 1.1816 \qquad \text{Referred to stator.}$$

$$\therefore \quad R_x = \frac{R_x}{k^2} = \frac{1.1816}{(2.5)^2} = 0.189\ \Omega/\text{ph} \qquad \text{Rotor side}$$

Example 1.37

A phase inductance motor runs at a speed of 1485 rpm at no load and at 1350 rpm at full load when supplied from a 50 Hz, 3-phase line.

(i) How many poles does the motor have?

(ii) What is the % slip at no-load and at full load.

(iii) What is the frequency of rotor voltage at no load and at full load?

(iv) What is the speed at both no-load and full load of the rotor field with respect to rotor conductor, the rotor field with respect to the stator and rotor field with respect to the stator field.

Solution: $N_{NL} = 1485$ rpm, $N_{FL} = 1350$ rpm $f = 50$- Hz.

(*i*) The nearest synchronous speed to N_{FL} is 1500 rpm

$$N_s = \frac{120f}{P} \quad i.e.,\ 1500 = \frac{120 \times 50}{P}$$

$\therefore \quad P = 4$

$$(ii) \quad S_{NL} = \frac{N_s - N_{NL}}{N_s} = \frac{1500 - 1485}{1500} = 0.01 \quad i.e.,\ 10\%$$

$$S_{FL} = \frac{N_s - N_{FL}}{N_s} = \frac{1500 - 1350}{1500} = 0.1 \text{ i.e., } 10\%$$

(*iii*) $(f_r)_{NL} = f \times S_{NL} = 50 \times 0.01 = 0.5$ Hz

$(f_r)_{FL} = f \times S_{FL} = 50 \times 0.1 = 5$ Hz

(*iv*) On no load

$$\text{Speed of rotor field with respect to rotor conductor} = \frac{120(f_r)_{NL}}{P} = \frac{120 \times 0.5}{4} = 15 \text{ rpm}$$

Rotor field with respect to stator = 1500 rpm

Rotor field with respect to stator field = 0 rpm

On full load

$$\text{Speed of rotor field with respect to rotor conductors} = \frac{120(f_r)_{FL}}{P} = \frac{120 \times 5}{4} = 150 \text{ rpm}$$

Rotor field with respect to stator = 1500 rpm

Rotor field with respect to stator field = 0 rpm

Example 1.38

A 3.3 kV, 20 pole, 50 Hz, 3-phase star connected IM has a slip ring rotor of resistance 0.025 ohm and standstill reactance of 0.28 ohm per phase. The motor has a speed of 294 rpm. When full-load torque is applied. Compute slip at maximum torque and the ratio of maximum to full load torque. Neglect stator impedance.

Solution: $V_L = 3.3$ kV, $P = 20$, $f = 50$ Hz, $R_2 = 0.025\ \Omega$, $X_2 = 0.28\ \Omega$

$N = 294$ rpm on full load

$$s_m = \frac{R_2}{X_2} = \frac{0.025}{0.28} = 0.0892 \text{ i.e., } 8.92\%$$

$$T \propto \frac{sE_2^2 R_2}{R_2^2 + (sX_2)^2}$$

$$\text{For full load } s = s_{fl} = \frac{N_s - N}{N_s}, \quad N_s = \frac{120 \times 50}{20} = 300 \text{ rpm}$$

$$\therefore \quad s_{fl} = \frac{300 - 294}{300} = 0.02 \text{ i.e., } 2\%$$

$\therefore$ For $T_m = s_m = 0.0892$

$$\therefore \quad \frac{T_m}{T_{F.L.}} = \frac{s_m}{R_2^2 + (s_m X_2)^2} \times \frac{R_2^2 + (s_{fl} X_2)^2}{s_{fl}} \qquad E_2, R_2 \text{ constants}$$

$$\frac{T_m}{T_{F.L.}} = \frac{0.0892}{[(0.025)^2 + (0.0892 \times 0.28)^2]} \times \frac{[(0.025)^2 + (0.02 \times 0.28)^2]}{0.02}$$

$$\frac{T_m}{T_{F.L.}} = 2.344$$

Example 1.39

The power input to a 3 phase 50 Hz induction motor is 50 kW. The total stator loss is 800 W. Find the total mechanical power developed, if the rotor emf makes 90 complete cycles per minute.

Solution: $P_{in} = 50$ kW, stator loss = 800 W, $f_r = 90$ cycles/min

$$f_r = 90 \text{ cycles/min} = \frac{90}{60} \text{ cycles/sec} = 1.5 \text{ Hz}$$

$f_r = sf$ *i.e.*, $1.5 = 5 \times 50$

$\therefore \quad s = 0.03$ *i.e.*, 3%

$P_2 = P_{in}$ – Stator loss = $50 \times 10^{-3} - 800 = 49200$ W.

Now $P_2 : P_c : P_m$ is $1 : s : 1 - s$

$$\therefore \qquad \frac{P_2}{P_m} = \frac{1}{1-s} \quad i.e., \quad \frac{49200}{P_m} = \frac{1}{(1-0.03)}$$

$$\therefore \qquad P_m = 47724 \text{ W}$$ Total mechanical power

Example 1.40

A 3-phase induction motor has a 4-pole, star connected stator winding. The motor runs on a 50 Hz supply with 200 V between lines. The rotor resistance and standstill rotor reactance per phase are 0.1 W and 0.9 Ω respectively. The ratio of rotor to stator turns in 0.67. Calculate:

(i) *Total torque at 4% slip.*

(ii) *Maximum torque developed.*

(iii) *Speed at maximum torque.*

(iv) *Maximum mechanical power.*

Neglect stator impedance.

Solution: $P = 4$, $V_L = 200$ V, $f = 50$ Hz, $R_2 = 0.1\ \Omega$, $X_2 = 0.9\ \Omega$, $s = 4\%$

$$K = \frac{E_{2\,ph}}{E_{1\,ph}} = 0.67, \quad E_{1\,ph} = \frac{V_2}{\sqrt{3}} = 115.47 \text{ V}$$

$$\therefore \qquad E_{2\,ph} = 0.67 \times 115.47 = 77.365 \text{ V} = E_2$$

$$N_s = \frac{120f}{P} = \frac{120 \times 50}{4} = 1500 \text{ rpm}$$

$$\therefore \qquad n_s = \frac{N_s}{60} = \frac{1500}{60} = 25 \text{ rps}$$

(i) $T = \dfrac{3}{2\pi n_s} \times \dfrac{sE_2^2 R_2}{R_2^2 + (sX_2)^2} = \dfrac{3}{2\pi \times 25} \times \dfrac{0.04 \times (77.365)^2 \times 0.1}{[(0.1)^2 + (0.04 \times 0.9)^2]}$

$= 40.4786$ N-m.

(ii) For maximum torque, $s = s_m = \dfrac{R_2}{X_2}$ hence,

$$\therefore \quad T_m = \frac{3}{2\pi n_s} \times \frac{E_2^2}{2X_2} = \frac{3}{2\pi \times 25} \times \frac{(77.365)^2}{2 \times 0.9} = 63.5065 \text{ N-m}$$

(iii) $s_m = \dfrac{R_2}{X_2} = \dfrac{0.1}{0.9} = 0.1111$

$\therefore \quad N_m = N_s(1 - s_m) = 1500\ (1 - 0.111) = 1333.333$ rpm

(iv) $P_{\max} = T_m \times w_m$

$$= T_m \times \frac{2\pi N_m}{60} = \frac{63.5065 \times 2 \times \pi \times 1333.333}{60} = 8867.1801 \text{ W.}$$

Example 1.41

A 6-pole, 50 Hz, 3-ϕ induction motor running on full load develops a useful torque of 150 N-m at a rotor frequency of 1.5 Hz. Calculate the shaft power output. If the mechanical torque lost in friction be 10 Nm. Determine

(i) Rotor copper loss.
(ii) The input to the motor.
(iii) The efficiency.

Solution: $P = 6$, $f = 50$ Hz, $T_{sh} = 120$ Nm, $f_r = 1.5$ Hz, $T_{\text{lost}} = 10$ Nm

$f_r = sf$ *i.e.,* $1.5 = s \times 50$

$\therefore \quad s = 0.03$ *i.e.,* 3%

$$N_s = \frac{120f}{P} = \frac{120 \times 50}{6} = 1000 \text{ rpm}$$

$$N = N_s(1 - s) = 1000(1 - 0.03) = 970 \text{ rpm}$$

$$\therefore \quad \text{Rotational loss} = T_{\text{lost}} \times w = 10 \times \left(\frac{2\pi \times 970}{60}\right) = 1015.7816 \text{ W}$$

$$P_{\text{out}} = T_{\text{sh}} \times w = 150 \times \left(\frac{2\pi \times 970}{60}\right) = 15236.7243 \text{ W}$$

$$P_m = P_{\text{out}} + \text{Rotational loss}$$

$$P_m = 15236.7243 + 1015.7816$$

$$P_m = 16252.5059 \text{ W}$$

(i) $P_2 : P_c : P_m$ *i.e.,* $1 : s : 1 - s$

$$\therefore \quad \frac{P_c}{P_m} = \frac{s}{1-s} \quad \text{i.e.,} \quad P_c = \frac{16252.5059 \times 0.03}{(1 - 0.03)}$$

$\therefore \quad P_c = 502.6548$ W Rotor copper loss

(ii) $\dfrac{P_2}{P_c} = \dfrac{1}{s}$ i.e., $P_2 = \dfrac{P_c}{s} = \dfrac{502.6548}{0.03} = 16755.16$ W

Assume total stator losses = 700 W

$$\therefore \quad P_{in} = P_2 + \text{Total stator losses} = 16755.16 + 700$$
$$= 17455.16 \text{ W}$$

(iii) $\%\eta = \dfrac{P_{out}}{P_{in}} \times 100 = \dfrac{15236.7243}{17455.16} \times 100 = 87.29\%$

STARTING AND SPEED CONTROL OF INDUCTION MOTORS

Example 1.42

A squirrel cage induction motor has a full load slip of 5%. The motor starting current at rated voltage is 6 times its full load current. Find the tapping on the autotransformer starter which would give full load torque at start. What would then be the supply starting current?

Solution: Starting current at rated voltage = I_{sc}

$\therefore \quad I_{sc} = 6I_{F.L.}$ and $s_f = 5\% = 0.05$

Let X = Tapping on autotransformer.

$$T_{F.L.} = T_{st}$$

$$\therefore \quad \frac{T_{st}}{T_{F.L.}} = X^2 \left[\frac{I_{sc}}{I_{F.L.}}\right]^2 \times s_f$$

$$\therefore \quad 1 = X^2 \left[\frac{6}{1}\right]^2 \times 0.05$$

$$\therefore \quad X = 0.7453$$

Now, $\quad I_{st}\text{ (supply)} = X\, I_{st}\text{ (motor)} = X[XI_{sc}] = X^2 I_{sc}$

$$= X^2 \times 6I_{F.L.} = 3.33\, I_{F.L.}$$

Thus supply starting current is 3.33 times the full load current.

Example 1.43

A three phase induction motor has a ratio of maximum torque to full load torque as 2 : 5 : 1. Determine the ratio of starting torque to full load torque if star-delta starter is used. The rotor resistance and standstill reactance per phase are 0.4 Ω and 4 Ω respectively.

Solution: The given ratio is, $\dfrac{T_m}{T_{F.L.}} = 2.5$

The rotor values are, $R_2 = 0.4\ \Omega$, $X_2 = 4\ \Omega$

Now, $$T_m = \frac{kE_2^2}{2X_2}$$

$\therefore$ $$T_{F.L.} = \frac{T_m}{2.5} = \frac{kE_2^2}{5X_2} = \frac{kE_2^2}{20}$$

Now $$T_{st} = \frac{k\,E_2^2 R_2}{R_2^2 + X_2^2}$$

with star-delta starter $E_2 = \dfrac{E_2}{\sqrt{3}}$

$\therefore$ $$T_{st} = \frac{k\left(E_2/\sqrt{3}\right)^2 R_2}{R_2^2 + X_2^2}$$

Taking ratio of equation (2) and (1)

$$\frac{T_{st}}{T_{F.L.}} = \frac{k\left(E_2/\sqrt{3}\right)^2 R_2}{R_2^2 + X_2^2} \times \frac{20}{kE_2^2} = \frac{20 \times 0.4}{3\left[(0.4)^2 + (4)\right]} = 0.165$$

Example 1.44

The outer and inner cages of a double cage induction motor have standstill impedances of (0.05 + j 0.10) Ω and (0.01 + j 0.6) Ω respectively. Find the ratio of the torque due to the two cages (i) at starting (ii) when running with 3% slip. Neglect stator impedance.

Solution: Given

$$Z_{\text{outer cage}} = (0.05 + j\,0.10)\ \Omega$$
$$Z_{\text{inner cage}} = (0.01 + j\,0.60\ \Omega$$
$\therefore$ $$R'_{20} = 0.05\ \Omega, \quad X'_{20} = 0.1\ \Omega$$
$$R'_{2i} = 0.01\ \Omega, \quad X'_{2i} = 0.6\ \Omega$$

Slip, $s = 3\% = 0.03$

We have to find the ratio of $T_{\text{outer cage}}$ and $T_{\text{inner cage}}$ at standstill.

The torque of induction motor is given by

$$T = \frac{3}{2\pi n_s} \times \left\{\frac{sE_2^2 R_2}{R_2^2 + (sX_2)^2}\right\}$$

(*i*) At starting $s = 1$

$$T = \frac{3}{2\pi n_s} \times \left\{\frac{E_2^2 R_2}{R_2^2 + X_2^2}\right\}$$

As, E_2, n_s are fixed.

$$T \propto \frac{R_2}{R_2^2 + X_2^2}$$

$$\therefore \quad \frac{T_{\text{outer cage}}}{T_{\text{inner cage}}} = \frac{R'_{20}\Big/\left(R'^2_{20} + X'^2_{20}\right)}{R'_{2i}\Big/\left(R'^2_{2i} + X'^2_{2i}\right)} = \frac{R'_{20}}{R'_{2i}} \frac{\left(R'^2_{2i} + X'^2_{2i}\right)}{\left(R'^2_{20} + X'^2_{20}\right)}$$

$$= \frac{0.05}{0.01} \frac{(0.01)^2 + (0.6)^2}{(0.05)^2 + (0.1)^2} = \frac{0.05}{0.01} \times \frac{0.3601}{0.0125} = 144$$

(*ii*) At $s = 3\%$ *i.e.*, $s = 0.03$

$$T \propto \frac{sR_2}{R_2^2 + (sX_2)^2} \quad \text{as } E_2, n_s \text{ are fixed}$$

$$\frac{T_{\text{outer cage}}}{T_{\text{inner cage}}} = \frac{\left[\dfrac{sR'_{20}}{R'^2_{20} + (sX'_{20})^2}\right]}{\left[\dfrac{sR'_{2i}}{R'^2_{2i} + (sX'_{2i})^2}\right]} = \frac{R'_{20}}{R'_{2i}} \frac{R'^2_{2i} + (sX'_{2i})^2}{R'^2_{20} + (sX'_{20})^2}$$

$$= \frac{0.05}{0.01} \frac{(0.01)^2 + (0.03 \times 0.6)^2}{(0.05)^2 + (0.03 \times 0.1)^2} = \frac{0.05}{0.01} \frac{4.24 \times 10^{-4}}{2.509 \times 10^{-3}}$$

$$\frac{T_{\text{outer cage}}}{T_{\text{inner cage}}} = 0.8449$$

Example 1.45

The cages of a double cage IM have standstill impedances of (3.5 + j 1.5) Ω and (0.6 + j 7) Ω. Full load slip is 6%. Find the starting torque in terms of full load torque. Neglect stator impedance and magnetising current.

Solution: The equivalent circuit of induction motor by neglecting stator impedance and magnetising current can be represented as,

Impedance of inner cage during starting is given as,

At starting $s = 1$

$$Z_i = R_i + jX_i = 0.6 + j\,7\ \Omega$$

Impedance of outer cage is

$$Z_0 = R_0 + jX_0 = 3.5 + j\,1.5\ \Omega$$

As inner and outer cages are in parallel,

$$\text{Equivalent impedance } Z = \frac{Z_i Z_0}{Z_i + Z_0} = \frac{(0.6 + j7)(3.5 + j1.5)}{(0.6 + j7)(3.5 + j1.5)}$$

$$= \frac{(7.025\angle 85.10°)(3.80\angle 23.19°)}{4.1 + j8.5}$$

$$= \frac{26.695\angle 180.29°}{9.437\angle 64.24°}$$

$$= 2.828\angle 44.05°\ \Omega = (2.032 + j\,1.9662)\ \Omega$$

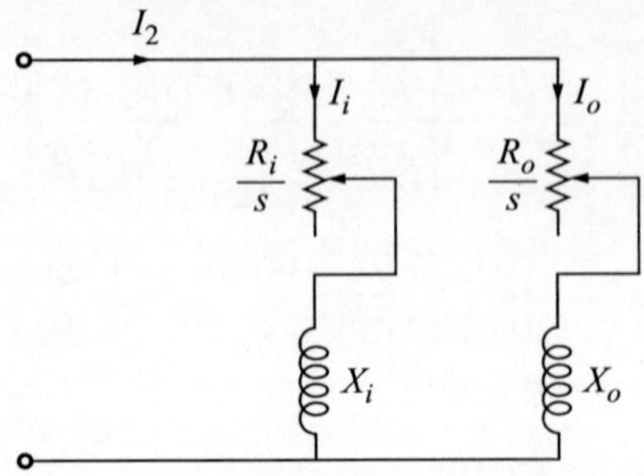

Fig. 1.38

$$\text{Combined resistance of rotor, } R_{eq} = 2.032 \text{ W}$$

$$\text{Rotor current } I_2 = \frac{V}{Z_{eq}} = \frac{V}{2.828}$$

Torque during starting in synchronous watts per phase is,

$$T_{st} = I_2^2 R_2 = \left(\frac{V}{2.828}\right)^2 (2.032)$$

During full load $s = 0.06$

$$Z_i = \frac{R_i}{s} + jX_i = \frac{0.6}{0.06} + j7 = 10 + j\,7\ \Omega$$

$$Z_0 = \frac{R_0}{s} + jX_0 = \frac{3.5}{0.06} + j1.5 = 58.33 + j\,1.5\ \Omega$$

$$\text{Equivalent impedance } Z = \frac{Z_i Z_0}{Z_i + Z_0} = \frac{(10 + j7)(58.33 + j1.5)}{(10 + j7)(58.33 + j1.5)}$$

$$Z = \frac{711.857 \angle 36.473°}{68.856 \angle 7.09} = 10.338 \angle 29.383°$$

$$Z = (9.0081 + j\,5.0722)\ \Omega$$

$$\text{Combined resistance } R_{eq} = 9.0081\ \Omega$$

$$\text{Rotor current } I_2 = \frac{V}{10.338}$$

$$T_{\text{full load}} = I_2^2 R_2 = \left(\frac{V}{10.338}\right)^2 \times 9.0081$$

$$\therefore \quad \frac{T_{st}}{T_{F.L.}} = \frac{\left[\frac{V^2}{(2.828)^2}\right](2.032)}{\left[\frac{V^2}{(10.338)^2}\right](9.0081)} = \frac{(10.338)^2 (2.032)}{(2.828)^2\ (9.0081)}$$

$$\therefore \quad T_{st} = (3 \times 100)\ T_{F.L.}$$

$$\therefore \quad T_{st} = 300\%\ T_{F.L.}$$

Example 1.46

A 4-pole, 50 Hz, 3-phase induction motor has rotor resistance of 0.2 Ω per phase and rotor standstill reactance of 1 Ω per phase. On full load it is running with a slip of 4%. Calculate the extra resistance required in the rotor circuit per phase to reduce the speed to 1260 rpm on the same load condition.

Solution: $P = 4, f = 50$ Hz, $R_2 = 0.2\ \Omega$ $X_2 = 1\ \Omega$, $s_f = 4\% = 0.04$

$$N_s = \frac{120f}{P} = \frac{120\times 50}{4} = 1500 \text{ rpm}$$

$$N_1 = \text{full load speed} = N_s(1 - s)$$

$$N_1 = 1500(1 - 0.04) = 1440 \text{ rpm}$$

$$N_2 = 1260 \text{ rpm}$$

Let new value of rotor resistance with extra resistance added by R_2' for achieving speed equal to N_2.

$\therefore$ $R_2' = R_2 + R_{ex}$ where R_{ex} = Extra resistance

Let $T = T_1$ for $N = N_1$ and $T = T_2$ for $N = N_2$

$$S_2 = \frac{N_s - N_2}{N_s} = \frac{1500-1260}{1500} = 0.16$$

Now $$T \propto \frac{sE_2^2R_2}{R_2^2 + (sX_2)^2}$$

Let $R_2 = R_2$ for $N = N_1$ and $R_2 = R_2'$ for $N = N_2$

$$\frac{T_1}{T_2} = \frac{s_f E_2^2 R_2}{R_2^2 + (s_f X_2)^2} \times \frac{[(R_2')^2 + (s_2 X_2)^2]}{s_2 E_2^2 R_2'}$$

But $T_1 = T_2$ as load is same

$$\therefore \quad s_f E_2^2 R_2[(R_2')^2 + (S_2X_2)^2] = s_2 E_2^2 R_2'[R_2^2 + (s_f X_2)^2]$$

$$0.04 \times 0.2[(R_2)^2 + (0.16 \times 1)^2] = 0.16 \times R_2'[(0.2)^2 + (0.04 + 1)^2]$$

$$R_2'^2 - 0.832R_2' + 0.0256 = 0$$

$$R_2' = \frac{0.832 \pm \sqrt{(0.832)^2 - 4\times 0.0256}}{2} = 0.032\ \Omega,\ 0.8\ \Omega$$

Now adding R_{ex}, R_2' cannot be less than R_2 so neglecting smaller value.

$$R_2' = 0.8 = R_2 + R_{ex}$$

$$R_{ex} = 0.8 - R_2 = 0.8 - 0.2 = 0.6\ \Omega \text{ per phase}$$

Example 1.47

A 3-phase, 4-pole, 415 V, 50 Hz induction motor has a star connected stator. The rotor impedance at standstill is 0.1 + j 0.9 Ω. The stator to rotor turns ratio is 1.75. Calculate the external resistance per phase required in the rotor to limit starting rotor current to 60 A, using rotor resistance starter.

Solution: $P = 4$, $V_L = 415$ V, $f = 50$ Hz, $E_2/E_1 = 1.75$

$R_2 = 0.1\ \Omega$, $X_2 = 0.9\ \Omega$, $I_2 = 60$ A at start

$$E_{1\text{ ph}} = \frac{V_2}{\sqrt{3}} = \frac{415}{\sqrt{3}} = 239.6 \text{ V} \qquad \text{star connection}$$

$$\frac{E_{2\text{ ph}}}{E_{1\text{ ph}}} = \frac{1}{1.75}$$

$\therefore$ $$E_{2\text{ ph}} = \frac{E_{1\text{ ph}}}{1.75} = \frac{239.6}{1.75} = 136.92 \text{ V}$$

Let $$R_2' = R_2 + R_{ex} \qquad \text{At start}$$

$\therefore$ $$I_2 = \frac{E_2}{\sqrt{(R_2')^2 + (X_2)^2}}$$

$\therefore$ $$60 = \frac{136.92}{\sqrt{(R_2')^2 + (0.9)^2}}$$

$\therefore$ $$(R_2')^2 + (0.9)^2 = \left(\frac{136.92}{60}\right)^2 = 5.2075$$

$\therefore$ $$(R_2')^2 = 4.3975$$

$\therefore$ $$R_2' = 2.097\ \Omega$$

But $$R_2' = R_2 + R_{ex}$$

$\therefore$ $$R_{ex} = R_2' - R_2 = 2.097 - 0.1 = 1.997\ \Omega \text{ per phase}$$

Example 1.48

A 50 Hz induction motor uses a pole amplitude modulation method to control the speed. The stator has 16 poles while the pole modulating function has 4 poles. At what two speeds motor can run?

Solution: $P = 16$ and $P_M = 4$

Let $n = 1$ and $r = 4$

Now $$\frac{n}{r} = \frac{1}{3}\left[1 \mp \frac{P_M}{P}\right]$$

$\therefore$ $$\frac{1}{3}\left[1 - \frac{P_M}{P}\right] = \frac{1}{3}\left[1 + \frac{4}{16}\right] = \frac{5}{12} \neq \frac{n}{r}$$

and $$\frac{1}{3}\left[1 + \frac{P_M}{P}\right] = \frac{1}{3}\left[1 + \frac{4}{16}\right] = \frac{5}{12} \neq \frac{n}{r}$$

Thus as first equation with –ve sign is satisfied, the P_1 poles are suppressed. So two possible speeds of motor are corresponding to

$P = 16$ and $P_2 = P + P_M = 20$

$\therefore$ $$N_s = \frac{120 f}{P} = \frac{120 \times 50}{16} = 375 \text{ rpm}$$

and $$N_s = \frac{120 f}{P_2} = \frac{120 \times 50}{20} = 300 \text{ rpm}$$

Example 1.49

Two induction motor are used for the cascade control. The main motor has 4 poles while the auxiliary motor has 6 poles. Determine the various synchronous speeds possible for the set. Assume supply frequency 50 Hz.

Solution: $P_A = 4$, $P_B = 6$ and $f = 50$ Hz

1. $N_s = \dfrac{120\,f}{P_A} = \dfrac{120 \times 50}{4} = 1500$ rpm. '*A*' running alone.

2. $N_s = \dfrac{120\,f}{P_B} = \dfrac{120 \times 50}{6} = 1000$ rpm '*B*' running alone.

3. $N_s = \dfrac{120\,f}{P_A + P_B} = \dfrac{120 \times 50}{4+6} = 600$ rpm Cumulative cascade

4. $N_s = \dfrac{120\,f}{P_A - P_B} = \dfrac{120 \times 50}{(4-6)} = -3000$ rpm Differential cascade

Though $P_A - P_B$ is negative, in practice the effective poles are 2 only. The negative sign indicates the direction of torque.

Example 1.50

A 4-pole induction motor and a 6 pole induction motor are connected in cumulative cascade arrangement. The supply frequency is 50 Hz while the frequency in rotor circuit of 4 poles motor is 1 Hz. Determine the slip of each machine and combined speed of the set.

Solution: The arrangement is shown in the Fig. 1.39.

f_{r1} = Rotor frequency of motor *A*

f_{r2} = Rotor frequency of motor *B* = 1 Hz

P_A = Poles of *A* = 4

P_B = Poles of *B* = 6

The set is cumulatively coupled.

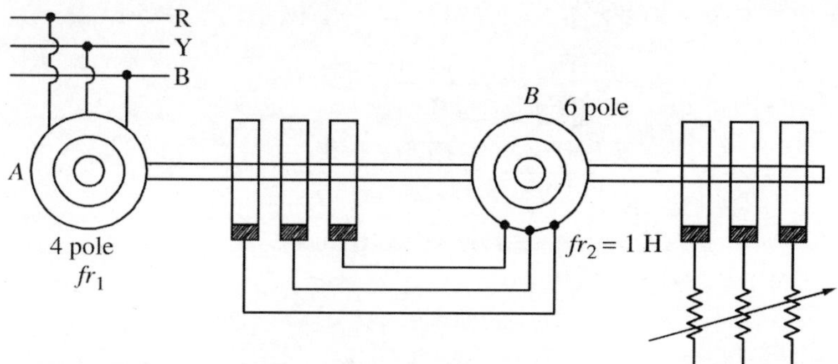

Fig. 1.39

Hence the synchronous speed of the set can be obtained as,

$$N_{sc} = \frac{120f}{P_A + P_B} = \frac{120 \times 50}{4+6} = 600 \text{ rpm}$$

Now, $$f_{r2} = sf$$

and $$s = \frac{N_{sc} - N}{N_{sc}} = \text{slip of the set}$$

i.e., $$1 = \frac{600 - N}{600} \times 50$$

$\therefore 12 = 600 - N$ *i.e.*, $N = 588$ rpm

This is the combined speed of the set

$$N_{SA} = \frac{120f}{P_A} = \frac{120 \times 50}{4} = 1500 \text{ rpm}$$

$$\therefore \quad s_A = \frac{1500 - 588}{1500} = 0.608 \text{ i.e., } 60.8\%$$

$$N_{SB} = \frac{120 f_{r1}}{P_B}$$

and $$f_{r1} = s_A f = 0.608 \times 50 = 30.4 \text{ Hz}$$

$$\therefore \quad N_{SB} = \frac{120 \times 30.4}{6} = 608 \text{ rpm}$$

$$\therefore \quad S_B = \frac{N_{SB} - N}{N_{SB}} = \frac{608 - 588}{608} = 0.033 \text{ i.e., } 3.3\%$$

The S_A and S_B are the slip values of machine *A* and *B* respectively.

Example 1.51

A 4-pole, 3 phase, 50 Hz, slip ring type induction motor has a rotor resistance of 0.25 Ω per phase and rotor reactance of 2 Ω per phase at standstill condition. It is running at 1455 rpm speed. Calculate the value of external resistance per phase required in the rotor circuit to reduce the speed by 17%. Assume load torque constant.

Solution: $P = 4$, $f = 50$ Hz, $R_2 = 0.25\ \Omega$ $X_2 = 2\ \Omega$, $N_1 = 1455$ rpm.

$$N_s = \frac{120f}{P} = \frac{120 \times 50}{4} = 1500 \text{ rpm}$$

$$S_1 = \frac{N_s - N_1}{N_s} = \frac{1500 - 1455}{1500} = 0.03$$

$$N_2 = N_1 - 0.17N_1 \text{ speed reduced by } 17\%$$

$$S_2 = \frac{N_s - N_2}{N_s} = \frac{1500 - 1207.65}{1500} = 0.1949$$

Now $$T \propto \frac{s E_2^2 R_2}{R_2^2 + (sX_2)^2}$$

As torque constant $T_1 = T_2$.

And for T_2, $R_2' =$ Rotor resistance $= R_2 + R_{ex}$.

$$\therefore \quad \frac{T_1}{T_2} = \frac{S_1 E_2^2 R_2}{[R_2^2 + (S_1 X_2)^2]} \times \frac{[(R_2')^2 + (S_2 X_2)^2]}{S_2 E_2^2 (R_2')} = 1$$

$$\therefore \quad S_2(R_2')\left[R_2^2 + (S_1 X_2)^2\right] = S_1 R_2[(R_2')^2 + (S_2 X_2)^2]$$

$$\therefore \quad 0.1949(R_2')\ [(0.25)^2 + (0.03 \times 2)^2] = 0.03(0.025) \times [(R_2')^2 + (0.1949 \times 2)^2]$$

$$\therefore \quad 0.0128(R_2') = 7.5 \times 10^{-3}\ (R_2'^2) + 1.1395 \times 10^{-3}$$

$$\therefore \quad (R_2')^2 - 1.7067\ R_2' + 0.1519 = 0$$

$$\therefore \quad R_2' = \frac{1.7067 \pm \sqrt{(1.7067)^2 4 - 4 \times 0.1519}}{2}$$

$$R_2' = 0.094\ \Omega,\ 1.6124\ \Omega$$

Neglecting lower value as less than R_2

$$\therefore \quad R_2' = R_2 + R_{ex} = 1.6124\ \Omega$$

$$R_{ex} = 1.6124 - 0.25 = 1.3625\ \Omega$$

This is the extra resistance per phase required in the rotor.

Example 1.52

A 3 phase, 6 pole, 50 Hz induction motor when fully loaded with a slip of 3%. Find the value of the resistance necessary in series per phase of the rotor to reduce the speed by 10%. Assume rotor resistance per phase is 0.2 Ω.

Solution: The given values are

$P = 6$, $f = 50$ Hz, $s_f = 3\% = 0.03$, $R_2 = 0.2\ \Omega$

$$N_s = \frac{120 f}{P} = \frac{120 \times 50}{60} = 1000 \text{ rpm}$$

$$N_1 = N_s(1 - s_f) = 1000(1 - 0.03) = 970 \text{ rpm}$$

The new speed is reduced by 10%

$$N_2 = 970 - (10\% \text{ of } 970) = 970 - 97 = 873 \text{ rpm}$$

$$S_2 = \frac{N_s - N_2}{N_s} = \frac{1000 - 873}{1000} = 0.127$$

Torque T is given as

$$T = \frac{k\,s\,E_2^2 R_2}{R_2^2 + (sX_2)^2}$$

if (sX_2) is neglected

$$T = \frac{k\,s\,E_2^2 R_2}{R_2^2} = \frac{k\,s\,E_2^2}{R_2}$$

where r is external resistance required to reduce the speed.

Now $T_{F.L.} = T_{20}$

$$\therefore \quad \frac{s_f}{R_2} = \frac{s_2}{R_2 + r} \quad i.e., \quad \frac{R_2 + r}{R_2} = \frac{s_2}{s_f}$$

$$\therefore \quad \frac{0.2 + r}{0.2} = \frac{0.127}{0.03} = 4.2333$$

$$\therefore \quad r = 0.65\ \Omega$$

Example 1.53

Determine approximately the starting torque of a three phase induction motor in terms of full load torque when started by (i) star-delta starter, (ii) an autotransformer with 50% tapping. The short circuit current of the motor is 5 times the full load current and full load slip is 5 percent.

Solution: The given values are

$I_{sc} = 5\ I_{F.L.}$ and $s_f = 5\% = 0.05$

(*i*) Using star-delta starter

$$\frac{T_{st}}{T_{F.L.}} = \frac{1}{3}\left(\frac{I_{sc}}{I_f}\right)^2 s_f = \frac{1}{3}(5)^2 \times 0.05 = 0.4166$$

$$T_{st} = 41.66\%\ T_{FL}$$

(*ii*) The autotransformer is having 50% tapping.

$$k = 0.5,\ k^2 = 0.25$$

$$\frac{T_{st}}{T_{FL}} = k^2\left(\frac{I_{sc}}{I_f}\right)^2 .s_f = (0.5)^2\ (5)^2\ (0.05) = 0.3125$$

$$\therefore \quad T_{st} = 31.25\%\ T_{FL}$$

Example 1.54

A 3 phase, 6 pole, 50 Hz induction motor takes 60 A at full load speed of 940 rpm develops a torque of 150 N-m. The starting current at rated voltage is 300 A. What is starting torque? If a star/delta starter is used determine the starting torque and starting current.

Solution: The given values are,

$I_{FL} = 60$ A, $P = 6$ pole $\quad N = 940$ rpm

$T_{FL} = 150$ Nm $\quad I_{sc} = 300$ A

We have
$$\frac{T_{st}}{T_{FL}} = \left(\frac{I_{sc}}{I_{FL}}\right)^2 s_f$$

Now,
$$N_s = \frac{120 f}{P} = \frac{120 \times 50}{60} = 1000 \text{ rpm}$$

$$s_f = \frac{N_s - N}{N_s} = \frac{1000 - 940}{1000} = 0.06$$

Substituting, $$\frac{T_{st}}{150} = \left(\frac{300}{60}\right)^2 (0.06) = 1.5$$

$$T_{st} = 1.5 \times 150 = 225 \text{ N-m}$$

If star-delta starter is used,

$$\frac{T_{st}}{T_{FL}} = \frac{1}{3}\left(\frac{I_{sc}}{I_f}\right)^2 s_f = \frac{1}{3}(1.5) = 0.5$$

$$T_{st} = 0.5T_{FL} = 0.5 \times 150 = 75 \text{ N-m}$$

To find starting current

$$I_{sc} = \sqrt{3\left(\frac{T_{st}}{T_{FL}}\right)\frac{1}{s_f} \cdot T_{FL}} = \sqrt{3\left(\frac{75}{150}\right)\left(\frac{1}{0.06}\right)} \cdot 60$$

$\therefore$ $$I_{sc} = 300 \text{ A}$$

Example 1.55

The ratio of maximum torque to full load torque in a three phase squirrel cage induction motor is 2.2 : 1. Determine the ratio of actual starting torque of full load torque for direct starting, star-delta starting and auto transformer starting with tapping of 70%. The rotor resistance and standstill reactance per phase are 0.5 Ω and 5 Ω respectively.

Solution: For T_{st}, $s = 1$ and $R_2 = 0.5\ \Omega$, $X_2 = 5\ \Omega$

$\therefore$ $$T_{st} \propto \frac{E_2^2 R_2}{R_2^2 + X_2^2} \text{ and } T_m \propto \frac{E_2^2}{2X_2} \qquad s_m = \frac{R_2}{X_2}$$

while $$\frac{T_m}{T_{FL}} = 2.2$$

$\therefore$ $$T_{FL} \propto \frac{1}{2.2}T_m = \frac{1}{2.2}\frac{E_2^2}{2X_2} = \frac{1}{4.4}\frac{E_2^2}{X_2}$$

$\therefore$ $$T_{FL} \propto \frac{E_2^2}{4.4 \times 5} \propto \frac{E_2^2}{22}$$

(*a*) For direct on line

$\therefore$ $$\frac{T_{st}}{T_{FL}} = \frac{E_2^2 R_2}{R_2^2 + X_2^2} \times \frac{22}{E_2^2} = \frac{0.5}{[(0.5)^2 + (5)^2]} \times 22 = 0.4356$$

(*b*) Star-delta starting

$$\text{Starting rotor voltage per phase} = \frac{E_2}{\sqrt{3}}$$

$$\therefore \quad \frac{T_{st}}{T_{FL}} = \frac{(E_2/\sqrt{3})^2 R_2}{R_2^2 + X_2^2} \times \frac{22}{E_2^2} = 0.1452$$

(*c*) Auto transformer starting,

Stator voltage per phase = 0.7 times the voltage on direct on line.

i.e., rotor voltage at starting = 0.7 E_2

$$\therefore \quad \frac{T_{st}}{T_{FL}} = \frac{(0.7\,E_2)^2 \,.\, R_2}{R_2^2 + X_2^2} \times \frac{22}{E_2^2} = (0.7)^2 \times 0.4356 = 0.2134$$

Example 1.56

A 3 phase squirrel cage induction motor has a maximum torque to thrice the full load torque. Determine the ratio of starting torque to full load torque if started by (i) DOL starter (ii) star-delta starter. The maximum torque occurs at 0.1 slip.

Solution: $T_m = 3\ T_{fl},\ s_m = 0.1$

(*i*) D.O.L. starter

$$\frac{T_{st}}{T_{fl}} = \frac{T_{st}}{T_m} \times \frac{T_m}{T_{fl}} = \frac{2S_m}{1+S_m^2} \times \left[\frac{T_m}{T_{fl}}\right] = \frac{2 \times 0.1}{[1+(0.1)^2]} \times 3$$

(*ii*) With star delta starter, the starting torque is one third of that obtained by D.O.L. starter.

$$\therefore \quad \frac{T_{st}}{T_{fl}} = \frac{1}{3} \times \left[\frac{T_{st}}{T_{fl}}\right]_{DOL} = \frac{1}{3} \times 0.594 = 0.198$$

Example 1.57

A 3 phase, 400 V distribution circuit is designed to supply 1200 A. Assuming that three phase squirrel cage induction motor has full load efficiency of 0.85 power factor 0.8 starting current is 5 times the rated current what is the maximum possible kW of motor if it designed to use star-delta starter?

Solution: Maximum possible permissible current that induction motor can takes from the distribution circuit is 1200 A at the time of starting.

But $\qquad I_{st} = 5 \times I_{rated}$

$$\therefore \quad I_{2\,(rated)} = \frac{I_{st}}{5} = \frac{1200}{5} = 240\ \text{A}$$

$$\therefore \quad \text{Maximum kW rating} = \sqrt{3} V_L I_L \cos\phi \times \eta$$

$$= \sqrt{3} \times 400 \times 240 \times 0.85 \times 0.8$$

$$= 113.068\ \text{kW}$$

If it is designed to have star-delta starter, then star-delta starter is equivalent to autotransformer with 57.8% tapping *i.e.*, ratio $(1/\sqrt{3})$.

$$\therefore \quad I_{st} = X_0^2 \times 5 \times I_L = \left(\frac{1}{\sqrt{3}}\right)^2 \times 5 \times I_L$$

$$\therefore \quad I_L = \frac{1200 \times 3}{5} = 720 \text{ A}$$

$$\therefore \quad (\text{KW})_{\max} = \sqrt{3} \times 400 \times 720 \times 0.8 \times 0.85 = 339.204 \text{ kW.}$$

Example 1.58

A cage type induction motor when started by means of a star-delta starter takes 180% of full load line current and develops 35% of full load torque at starting. Calculate the starting torque and current in terms of full load values, if an autotransformer with 80% tappings were employed.

Solution: (*a*) Star-delta starting

$I_{sc} = 3 \times 180\%$ of $I_{fl} = 3 \times 1.8\ I_{fl} = 5.4\ I_{fl}$

$$\therefore \quad \frac{T_{st}}{T_{fl}} = \frac{1}{3}\left[\frac{I_{sc}}{I_{fl}}\right]^2 S_{fl} \quad \text{and} \quad T_{st} = 0.35\ T_{fl} \text{ (given)}$$

$$\therefore \quad 0.35 = \frac{1}{3} \times (5.4)^2 \times S_{fl}$$

$$\therefore \quad S_{fl} = 0.036 \textit{ i.e.}, 3.6\%$$

(*b*) Autotransformer starting with $x = 0.8$ *i.e.*, 80% tapping.

$$\therefore \quad I_{st} = X^2 I_{sc} = (0.8)^2 I_{sc} = 0.64\ I_{sc} = 0.64 \times 5.4\ I_{fl}$$

$$\therefore \quad I_{st} = 3.456\ I_{fl}$$

$$\frac{T_{st}}{T_{fl}} = X^2\left[\frac{I_{sc}}{I_{fl}}\right]^2 S_{fl} = (0.8)^2 \times (5.4)^2 \times 0.036 = 0.6718$$

$$\therefore \quad T_{st} = 0.6718\ T_{fl} \textit{ i.e.}, 67.18\% \text{ of full load torque.}$$

Example 1.59

The impedance at standstill of the inner and outer windings of a double cage rotor are 0.015 + 0.5j Ω and 0.05 + j 0.11 Ω respectively. Calculate the ratio of torque due to the two windings at standstill.

Solution: $Z_{\text{outer}} = 0.05 + j\ 0.11\ \Omega$, $Z_{\text{inner}} = 0.015 + j\ 0.5\ \Omega$

$\therefore\ R'_{20} = 0.05\ \Omega$, $X'_{20} = 0.11\ \Omega$, $R'_{2i} = 0.15\ \Omega$, $X'_{2i} = 0.5\ \Omega$

For an induction motor, $T = \dfrac{3}{2\pi n_s} \times \dfrac{s E_2^2 R_2}{R_2^2 + (sX_2)^2}$

At standstill, $s = 1$ and n_s, E_2 are constants,

$$\therefore \quad T \propto \frac{R_2}{R_2^2 + X_2^2}$$

$$\therefore \quad \frac{T_{outer}}{T_{inner}} = \frac{R'_{20}}{(R'_{20})^2 + (X'_{20})^2} \times \frac{(R'_{2i})^2 + (X'_{2i})^2}{R'_{2i}}$$

$$= \frac{0.05\,[(0.015)^2 + (0.5)^2]}{0.015[(0.05)^2 + (0.11)^2]} = 57.13$$

Example 1.60

Two 3 phase slip ring induction motor each having 4 poles operates in cascade connection. The supply is 440 volt 50 Hz. What are the obtainable speed? Give suitable reason for speed during differential cascade.

Solution: $P_A = P_B = 4, f = 50$ Hz.

In cascade control following synchronous speeds can be achieved:

(*a*) Independently with A, $N_s = \dfrac{120\,f}{P_A} = 1500$ rpm.

(*b*) Independently with B, $N_s = \dfrac{120\,f}{P_B} = 1500$ rpm.

(*c*) Running as cumulatively cascaded,

$$N_s = \frac{120\,f}{P_A + P_B} = \frac{120 \times 50}{(4+4)} = 750 \text{ rpm}$$

(*d*) Running as differentially cascaded

$$N_s = \frac{120\,f}{P_A - P_B} = \frac{120 \times 50}{(4-4)} = \text{undefined}$$

Infect as the two poles P_A and P_B are same, only two speeds are possible. In differentially cascade connection the two rotating magnetic fields rotate in opposite direction with same speed and hence effectively the set will not rotate at all. Practically the motors with same number of poles are not used in cascade control.

Example 1.61

A cage induction motor when started by means of a star-delta starter takes 180% of full load line current k develops 35% of full load torque at starting. Calculate the starting torque and current in terms of full-load values, if an auto-transformer with 75% tapping were employed.

Solution: (*a*) Start-delta starting

$I_{sc} = 3 \times 180\%$ of $I_{fl} = 3 \times 1.8\, I_{fl} = 5.4\, I_{fl}$

$$\therefore \quad \frac{T_{st}}{T_{fl}} = \frac{1}{3}\left[\frac{I_{sc}}{I_{fl}}\right] S_{fl} \text{ and } T_{st} = 0.35\ T_{fl} \text{ (given)}$$

$$\therefore \quad 0.35 = \frac{1}{3} \times (5.4)^2 \times S_{fl}$$

$$\therefore \quad S_{fl} = 0.036 \quad i.e., \quad 3.6\%$$

(*b*) Autotransformer starting with $X = 0.75$ *i.e.*, 75% tapping

$$\therefore \quad I_{st} = X^2 I_{sc}$$

$$\therefore \quad I_{st} = (0.75)^2 I_{sc} = 0.5625\ I_{sc}$$

$$\therefore \quad I_{st} = 0.5625 \times 5.4\ I_{fl}$$

$$\frac{T_{st}}{T_{fl}} = X^2\left[\frac{I_{sc}}{I_{fl}}\right]^2 S_{fl} = (0.75)^2 \times (5.4)^2 \times 0.036 = 0.59049$$

$$\frac{T_{st}}{T_{fl}} = 0.59049$$

$$\therefore \quad T_{st} = 0.59049\ T_{fl} \quad i.e., \quad 59.49\% \text{ of full load torque.}$$

Example 1.62

A 3-phase, delta-connected cage type induction motor when connected directly to 400 V, 50 Hz supply takes a starting current of 100 A in each stator phase. Calculate

(*i*) The line current for direct-on-line starting.

(*ii*) Line and phase starting currents for star-delta starting.

(*iii*) Line and phase starting currents for a 70% tapping on auto-transformer starting.

Solution: (*i*) In D.O.L. starter

$$I_L = \sqrt{3} I_{ph} = \sqrt{3} \times 100 = 173.205 \text{ A}$$

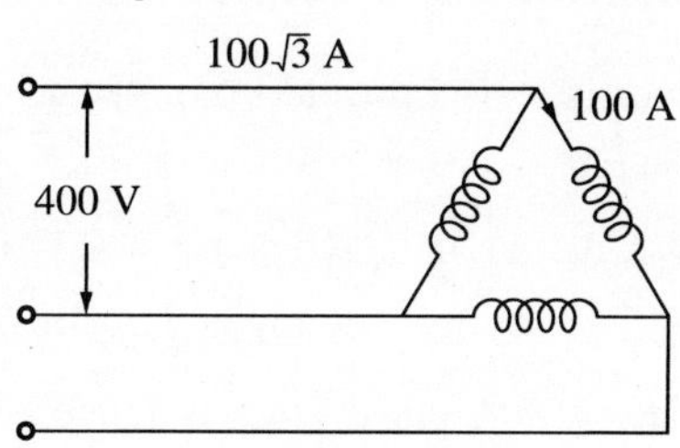

Fig. 1.40

(*ii*) In star-delta starter

$$V_{ph} = \frac{V_L}{\sqrt{3}} = \frac{400}{\sqrt{3}} = 230.94 \text{ V}$$

This will produce a phase current of, $I_{ph} = \dfrac{100}{\sqrt{3}} = 57.735$ A.

This is because voltage is reduced by $1/\sqrt{3}$, hence current will also reduce by $1/\sqrt{3}$.

$\therefore$ Starting phase current = 57.735 A in start connection, $I_L = I_{ph}$

$\therefore$ Starting line current = 57.735 A

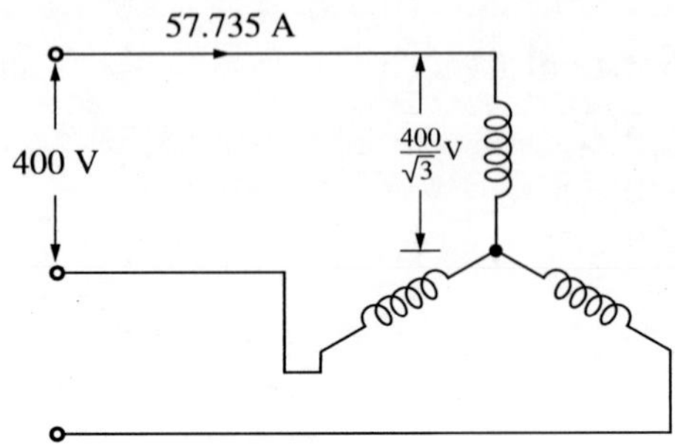

Fig. 1.41

(*iii*) 70% tapping on autotransformer starting.

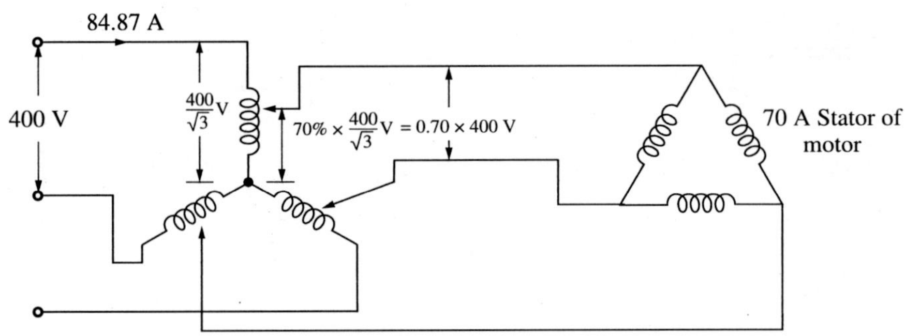

Fig. 1.42 Autotransformer

The autotransformer is star connected. So, phase voltage of autotransformer is $400/\sqrt{3} = 230.94$ V.

The line voltage for delta connected stator is 0.7 × 400 = 280 V.

$\therefore$ The phase voltage for delta connected stator is 280 V.

The 400 V circulates 100 A phase current hence 280 V will circulate a phase of $\frac{280}{400} \times 100 = 70$ A.

$\therefore$ Starting phase current of motor = 70 A

$\therefore$ Starting line current of motor = $\sqrt{3} \times 70 = 121.243$ A

Supply line current = 0.7 × 121.243 = 84.87 A.

SOLVED PROBLEMS

Example 1.63

A, 12 pole, 3-phase alternator is coupled to an engine running at 500 rpm. It supplies an induction motor which has a full-load speed of 1440 rpm find the slip and the number of poles of the motor.

Solution:
$$f_1 = \frac{PN_s}{120}, \frac{12 \times 500}{120} = 50 \text{ Hz}$$

The speed of the induction motor is 1440 rpm. Under normal operating conditions, an induction motor operates at a speed slightly less than is synchronous speed. The supply frequency for induction motor is 50 Hz. The possible synchronous speeds for 50 Hz are 3000, 1500, 1000 rpm, etc.

So, the closed synchronous speed corresponding to the actual speed of 1440 rpm is 1500 rpm that is, N_{sm} = 1500 rpm

Slip of the motor

$$S = \frac{N_{sm} - N_r}{S_{sm}} = \frac{1500 - 1440}{1500} = 0.04 = 4\%$$

Number of poles of the motor

$$P_m = \frac{120 f}{N_{sm}} = \frac{120 \times 50}{1500} = 4$$

Example 1.64

The frequency of the emf in the stator of a 4 pole induction motor is 50 Hz, and that in the rotor is 1.5 Hz. What is the slip, and of what speed in the motor running.

Solution:
$$f_2 = sf_1$$

$$S = \frac{f_2}{f_1} = \frac{1.5}{50} = 0.03 \text{ p.u.} = 3\%$$

$$f_1 = \frac{PN_s}{120}$$

$$N_s = \frac{120 f_1}{P} = \frac{120 \times 50}{4} = 1500 \text{ rpm}$$

Speed of the motor

$$N = (1 - S)N_s = (1 - 0.03) \times 1500 = 1455 \text{ rpm}$$

Example 1.65

A 3-phase, 6-pole, 50 Hz induction motor has a slip of 1% at no load, and 3% at full load. Determine

(a) Synchronous speed (b) no-load (c) full load speed (d) frequency of rotor current at standstill (e) frequency of rotor current of full load.

Solution:

(*a*) $N_s = \dfrac{120 f_1}{P} = \dfrac{120 \times 50}{6} = 1000$ rpm

(*b*) $N_0 = (1 - S_0)\, N_s = (1 - 0.01) \times 1000 = 990$ rpm

(*c*) $N_{fl} = (1 - S_{fl})\, N_s = (1 - 0.03) \times 1000 = 930$ rpm

(*d*) Frequency of rotor current of standstill

$$f_2 = sf_1 = 1 \times 50 = 50 \text{ Hz}$$

(*e*) Frequency of rotor current at full load

$$f_2 = S_{fl} f_1 = 0.03 \times 50 = 1.5 \text{ Hz}$$

Example 1.66

A 3-phase, 50 Hz, 4-pole induction motor has a slip of 4%. Calculate (a) speed of the motor, (b) frequency of rotor emf.

If the rotor has a resistance of 1 Ω and standstill reactance of 4 Ω. Calculate the power factor (i) at standstill (ii) of a speed of 1400 rpm.

Solution:

$$N_s = \frac{120 f_1}{P} = \frac{120 \times 50}{4} = 1500 \text{ rpm}$$

$$S = 4\% = 0.04 \text{ pu}$$

(*a*) Speed of the motor

$$N = (1 - S)N_s = (1 - 0.04) \times 1500 = 1440 \text{ rpm}$$

(*b*) Frequency of rotor emf

$$f_2 = sf_1 = 0.04 \times 50 = 2 \text{ Hz}$$

(*i*) $R_2 = 1\ \Omega,\ X_2 = 4\ \Omega$

Rotor impedance at standstill

$$Z_2 = R_2 + jX_2 = 1 + j4 = 4.123 \angle 75.96^\circ\ \Omega$$

Power factor at standstill

$$\text{Cos } \phi_2 = \cos 75.96^\circ = 0.2425 \text{ (log)}$$

(*ii*) The slip of a speed of 1400 rpm is

$$S_1 = \frac{N_s - N}{N_s} = \frac{1500 - 1400}{1500} = \frac{1}{15}$$

Rotor impedance of slip S_1 is

$$Z_{2s1} = R_2 + jS_1X_2$$

$$= 1 + j \times \frac{1}{0.5} X_4 = 1 + j\, 0.2667 = 1.03495 \angle 14.93^\circ\ \Omega$$

Power factor at 1400 rpm is

$$\cos \phi_{2s1} = \cos 14.93^\circ = 0.9662 \text{ (lag)}$$

Example 1.67

A 3-phase slip-ring induction motor gives a reading of 60 V across slip rings when at rest with normal stator voltage applied. The rotor is star-connected and has an impedance of (0.8 + j6) Ω per phase. Find the rotor current when the machine is

(a) at standstill with the slip-rings joined to a star-connected stator with a phase impedance of (4 + j3) Ω and (b) running normally with a 5% slip.

Solution: E_2 = emf induced per phase of the rotor of standstill

$$= \frac{60}{\sqrt{3}} = 34.64 \text{ V}$$

Total impedance of the rotor at stand still = impedance of rotor + impedance of state starter

$$= (0.8 + j6) + (4 + j3) = 4.8 + j9\ \Omega$$

(*a*) Current at standstill

$$I_2 = \frac{E_2}{Z_2} = \frac{32.64\angle 0°}{4.8 + j9}$$

$$= \frac{34.64\angle 0°}{10.2\angle 61.93°} = 3.396\angle -61.93° \text{ A}$$

(*b*) $S = 5\% = 0.05$ p.u.

During normal running, the starting resistance are cut-off.

$$I_{2s} = \frac{E_{2s}}{Z_{2s}} = \frac{sE_2}{R_2 + jX_2}$$

$$= \frac{0.05 \times 34.64}{0.8 + j0.05 \times 6} = \frac{1.732}{0.8 + j\,0.3} = \frac{1.732}{0.8544\angle 20.56°}$$

$$= 20.27\angle -20.56° \text{ A}$$

Example 1.68

The power input to a 3-phase including motor is 60 kW. The stator laser total 1 kW. Find the total mechanical power developed and the rotor copper loss per phase if the motor is running with a slip of 3%.

Solution: Stator input $P_{is} = 60$ kW, $S = 3\% = \frac{3}{100} = 0.03$ pu

$$\begin{aligned}
\text{Stator lasers} &= 1 \text{ kW} \\
\text{Stator output} &= 60 - 1 = 59 \text{ kW} \\
\text{Rotor input} &= \text{Stator output} = 59 \text{ kW} \\
\text{Total rotor copper loss} &= S \times \text{rotor input} = 0.03 \times 59 = 1.77 \text{ kW} \\
\text{Rotor copper loss per phase} &= \frac{1}{3} \times 1.77 = 0.59 \text{ kW} \\
\text{Mechanical power developed} &= \text{rotor input} - \text{rotor copper loss} \\
&= 59 - 1.77 = 57.23 \text{ kW}
\end{aligned}$$

Example 1.69

A, 6-pole, 50 Hz, 3-ϕ induction motor running on full load develops a useful torque of 150 Nm at a rotor frequency of 1.5 Hz. Calculate the shaft power output. If the mechanical torque lost in friction be 20 Nm. Determine (a) rotor copper loss, (b) the input to the motor (c) the efficiency, the total stator loss in 700 W.

Solution:

$$N_s = \frac{120 f_1}{P} = \frac{120 \times 50}{6} = 1000 \text{ rpm}$$

$$S = \frac{f_2}{f_1} = \frac{1.5}{50} = 0.03 \text{ or } 3\%$$

$$N_r = (1 - S)\, N_s = (1 - 0.003) \times 1000 = 970 \text{ rpm}$$

$$w_r = 2\pi n_r = \frac{2\pi \times 970}{60} = 101.58 \text{ rad/s}$$

$$\text{Shaft power output, } P_0 = Z_0 w_r$$

$$= 150 \times 101.58 = 15236 \text{ W} = 15.236 \text{ kW}$$

Mechanical power developed

$$P_{md} = (150 + 10) \times 101.58 = 16252 \text{ W} = 16.252 \text{ kW}$$

(*a*) Rotor copper loss $P_{rc} = \left(\frac{S}{1-S}\right) P_{md}$

$$= \frac{6.03}{1-0.03} \times 16252 = 502.6 \text{ W} = 0.5026 \text{ kW}$$

(*b*) Input to motor, $P_i = P_{md} + P_{rc} + P_{sc}$

$$= 16.252 + 0.5026 + 0.700 = 17.4546 \text{ kW}$$

(*c*) Efficiency $= \frac{P_0}{P_i} = \frac{15.236}{17.4546} = 0.8727 \text{ pu} = 87.29\%$

Example 1.70

The power input to the rotor of 440 V, 50 Hz, 6-pole, 3-phase induction motor is 80 kW. The rotor emf is observed to make 100 complete other nations per min. Calculate

(a) the slip (b) the rotor speed (c) the mechanical power developed (d) the rotor copper loss per phase (e) the rotor resistance per phase if the rotor current is 65 A.

Solution: $f_1 = 50 \text{ Hz}, f_2 = \frac{100}{60} \text{ Hz}$

$$f_2 = sf_1,\ s = \frac{f_2}{f_1} = \frac{100}{60 \times 50} = 0.033 \text{ pu}$$

(*a*) $S = \frac{N_s - N_r}{N_s},\ N_s = \frac{120 f}{P} = \frac{120 \times 50}{6} = 1000 \text{ rpm}$

(*b*) $N_r = (1 - S) = \left(1 - \frac{2}{60}\right) \times 1000 = 1000 - 33.3 = 966.7 \text{ rpm}$

(*c*) Mechanical power developed = rotor input – rotor copper loss

Also rotor copper loss = 5 × rotor input

$$= \frac{1}{30} \times 80 \times 1000 \text{ W}$$

$$\text{Mechanical power developed} = 80 \times 1000 - 2667 = 77333 \text{ W}$$

$$= \frac{77333}{746} h_p = 103.66\, n_p$$

$$(d) \text{ Rotor copper loss per phase} = \frac{80 \times 1000}{30 \times 3} = 889 \text{ W}$$

$$(e) \text{ Rotor resistance per phase } R_2 = \frac{\text{rotor per loss/phase}}{I^2} = \frac{889}{652} = 0.2104\ \Omega$$

Example 1.71

A 25 hp 6-pole, 50 Hz, 3-phase slip-ring induction motor runs at 960 revolutions per minute on full load with a rotor. Current per phase of 35 A. Allowing 250 W for the copper loss in the short-circuiting gear, and 1000 W by mechanical losses. Find the resistance per phase of the three-phase rotor winding.

Solution:

$$f_1 = \frac{PN_s}{120}$$

$$50 = \frac{6 \times N_s}{120}, \quad N_s = \frac{120 \times 50}{6} = 1000 \text{ rpm}$$

$$S = \frac{N_s - N_r}{N_s} = \frac{1000 - 960}{1000} = 0.04 \text{ pu}$$

Rotor copper loss

$$= \frac{S}{1 - S} \times \text{mechanical power developed}$$

$$3I_2^2 R_2 + 250 = \frac{0.04}{1 - 0.04} (25 + 746 + 1000)$$

$$3 \times 35^2 R_2 = 818.75 - 250$$

$$R_2 = \frac{568.75}{3 \times 35^2} = 0.15476\ \Omega$$

Example 1.72

A 500 V, 6-pole, 50 Hz, 3-ϕ induction motor develops 20 kW inclusive of mechanical laser when running of 995 rpm, the pf being 0.87. Calculate (a) the slip, (b) the rotor I^2R loss, (c) the total input of the stator loss is 1500 W, (d) line current (e) the rotor current frequency.

Solution:

$$N_s = \frac{120 f_1}{P} = \frac{120 \times 50}{6} = 1000 \text{ rpm}$$

$$S = \frac{N_s - N_r}{N_s} = \frac{1000 - 995}{1000} = 0.005 \text{ pu}$$

$$\text{Rotor on loss} = \text{slip} \times \text{rotor power input}$$
$$= S(\text{mech power developed} + \text{rotor cu loss})$$
$$P_{rc} = S(P_m + P_{rc})$$
$$P_{rc}(1 - S) = sP_m$$
$$P_{rc} = \frac{SP_m}{(1-S)} = \frac{0.005}{1-0.005} \times 20 \times 1000 = 100.5 \text{ W}$$

$$\text{Total input to stator} = \text{rotor power input} + \text{stator loss}$$

$$\text{Rotor input} = \frac{1}{S} \times \text{rotor cu loss}$$
$$= \frac{1}{0.005} \times 100.5 = 20100 \text{ W} = 20.1 \text{ kW}$$

$$\text{Hence total input} = 20.1 + 1.5 = 21.6 \text{ kW}$$

(*d*) $$\text{Line current} = \frac{21600}{\sqrt{3} \times 500 \times 0.87} = 28.7 \text{ A}$$

(*e*) $$f_r = Sf_1 0.005 \times 50 = 0.25 \text{ Hz}$$

Example 1.73

The stator loss of a 3-ϕ including motor is 2 kW. When the power input is 90 kW. What will be the rotor mechanical power developed and the rotor copper loss if the motor is running with a slip of 4%.

Solution: $S = 0.04$, stator input = 90 kW, stator loss = 2 kW

Rotor input P_{ri} = stator output

= stator input – stator loss = 90.2 = 80 kW

Rotor cu loss $P_{c2} = S \times$ rotor input $= 0.04 \times 88 = 3.52$ kW

Mechanical power developed

$$P_m = P_{ri} - P_{cr}$$
$$= 88 - 3.52 = 84.48 \text{ kW}$$

Example 1.74

A 3-phase induction motor with star-connected rotor has an induced emf of 60 V between slip right of standstill on open circuit with normal village applied to the stator. The resistance and standstill reactance of each rotor phase are 0.6 Ω and 4 Ω respectively. Calculate the currents per phase in the rotor:

(*a*) *when at standstill and connected to a star-connected rheostat of resistance 5 Ω and reactance 2 Ω per phase.*
(*b*) *when running short-circuited with 4% slip.*

Solution: (*a*) $E_2 = \dfrac{60}{\sqrt{3}}$ V per phase

Total rotor resistance = 0.6 + 5 = 5.6 Ω

Total rotor reactance = 4 + 2 = 6 Ω

$$I_2 = \frac{E_2}{Z_2} = \frac{60\sqrt{3}}{\sqrt{R_2^2 + X_L^2}} = \frac{60/\sqrt{3}}{\sqrt{(5.6)^2 + 6^2}} = 4.22 \text{ A}$$

(b) $E_{2s} = sE_2 = \frac{0.04 \times 60}{\sqrt{3}} = 1.3856$ V

$$Z_{2s} = \sqrt{R_2^2 \times (sX_2)^2} = \sqrt{(0.6)^2 + (0.04 \times 4)^2} = 0.62\ \Omega$$

$$I_{2s} = \frac{E_{2s}}{Z_{2s}} = \frac{1.3856}{0.62} = 2.235 \text{ A}$$

Example 1.75

The rotor resistance and standstill reactance per phase of a 3-ϕ slip-ring induction. Motor are 0.05 Ω and 0.2 Ω respectively. What should be the value of external resistance per phase to be inserted in the rotor circuit to give maximum torque of starting?

Solution: External resistance per phase added to the rotor circuit be *r* ohms

Rotor resistance per phase, $R_2 = (0.05 + \Omega)$

The starting torque will be maximum when

$$R_2 = X_2$$

$$0.05 + \Omega = 0.1,\ \Omega = 0.05\ \Omega$$

Example 1.76

A 6-pole, 3ϕ, 50 Hz induction motor runs on full load with a slip of 4 per cent. Given the rotor standstill impedance per phase as (0.01 + j 0.05) Ω. Calculate the available maximum torque in terms of full-load torque. Also determine the speed of which the maximum torque occurs.

Solution: $S = 4\% = 0.04$

$$S_m = \frac{R_2}{X_2} = \frac{0.01}{0.05} = 0.2$$

$$N_m = (1 - S_m)\, N_s = (1 - 0.2) \times 1000 = 800 \text{ rpm}$$

$$\frac{\tau_{max}}{\tau_{fl}} = \frac{S^2 + S_m^2}{2s\,S_m} = \frac{(0.04)^2 + (0.2)^2}{2 \times 0.04 \times 0.2} = 2.6$$

$$\therefore \quad \tau_{max} = 2.6\, T_{fl}$$

Example 1.77

A 6-pole, 3-phase, 50 Hz, induction motor develops a maximum torque of 30 Nm at 960 rpm. Determine the torque exerted by the motor at 5% slip. The rotor resistance per phase is 0.6 Ω.

Solution: $$N_s = \frac{120f}{P} = \frac{120\times 50}{6} = 1000 \text{ rpm}$$

Speed of maximum torque, $N_m = 960$ rpm

Slip at maximum speed

$$S_m = \frac{N_s - N_m}{N_s} = \frac{1000-960}{1000} = 0.04$$

Also $$S_m = \frac{R_2}{X_2} = X_2 = \frac{R_2}{S_m} = \frac{0.6}{0.04} = 15\ \Omega$$

If T_s is the torque of slip S

$$\frac{\tau_s}{\tau_{max}} = \frac{2s S_m}{S^2 + S_m^2}$$

Here $S = 0.05$, $\tau_{max} = 30$ Nm

$$\tau_s = \frac{2\times 0.05\times 0.04}{(0.05)^2 + (0.04)^2}\times 30 = 29.27 \text{ Nm}$$

Example 1.78

A 4-pole, 50 Hz, 3-phase induction motor develops a maximum torque of 110 Nm of 1360 rpm. The resistance of the star-connected rotor is 0.25 Ω/phase. Calculate the value of resistance that must be inserted in series, with each rotor phase to produce at starting torque equal to half the maximum torque.

Solution: Synchronous speed, $N_s = \frac{120f}{P} = \frac{120\times 50}{4}$

Speed at maximum torque $N_m = 1360$ rpm

$$\text{Slip at maximum torque } S_m = \frac{N_s - N_m}{N_s}$$

$$= \frac{1500-1360}{1500} = 0.0933$$

Also $$S_m = \frac{R_2}{X_2}$$

$$X_2 = \frac{R_2}{S_m} = \frac{0.25}{0.0933} = 2.68\ \Omega$$

$$T_{max} = \frac{kE_2^2}{2X_2} = \frac{k}{2X_2} = \frac{k}{2\times 2.68} = 0.1866 \text{ k}$$

where $k = kE_2^2$.

where r be the external resistance inserted per phase in the rotor circuit then starting torque.

$$\tau_{st} = \frac{kE_2^2(R_2 + r)}{(R_2 + r)^2 + X_2^2} = \frac{k(0.25 + r)}{(0.25 + r)^2 + (2.68)^2}$$

Let $\quad 0.25 + r = R_T$

$$\therefore \quad \tau_{st} = \frac{kR_T}{R_T^2 + (2.68)^2}$$

It is given that

$$\tau_{st} = \frac{1}{2}\tau_{max}$$

$$\frac{kR_T}{R_T^2 + (2.68)^2} = \frac{1}{2} \times 0.1866 \text{ k}$$

$$R_T^2 + (2.68)^2 = \frac{2R_T}{0.1866}$$

$$R_T^2 - 10.718\, R_T + 7.1824 = 0$$

$$R_T = 10.718 \pm \frac{\sqrt{(10.718)^2 - 4 \times 7.1824}}{2}$$

$$= \frac{1}{2}(10.718 \pm 9.28) = 9.999\ \Omega \text{ or } 0.7\ \Omega$$

$$R_T = 0.25 + r$$

$$r = R_T - 0.25 = 9.999 - 0.25 = 9.749\ \Omega$$

or $\quad r = 0.719 - 0.25 = 0.469\ \Omega$

We regent the value $r = 9.749\ \Omega$ as it corresponds to τ_{max} lying in the region where $S > 1$

$$r = 0.469\ \Omega$$

Example 1.79

A 746 kW, 3-phase, 50 Hz, 16 pole induction motor has a rotor impedance of (0.02 + j0.15) Ω at standstill, full load torque is obtained at 360 rpm. Calculate

(a) *the speed at which maximum torque occurs.*

(b) *the ratio of maximum to full-load torque.*

(c) *the external resistance per phase to be inserted in the rotor circuit to get maximum torque at starting.*

Solution: $\quad N_s = \frac{120f}{P} = \frac{120 \times 50}{16}$

$$\text{Speed at full load} = 360° \text{ rpm}$$

$$\text{Slip at full load } S_{fl} = \frac{N_s - N_r}{N_r} = \frac{375 - 360}{375} = 0.04$$

Slip at maximum torque

$$S_m = \frac{R_2}{X_2} = \frac{0.02}{0.15} = \frac{2}{15}$$

(*a*) Speed at which maximum torque occurs.

$$N_m = (1 - S_m)N_s = \left(1 - \frac{2}{15}\right) \times 375 = 325 \text{ rpm}$$

$$(b)\ \frac{\tau_{max}}{\tau_{fl}} = \frac{S_{fl}^2 + S_m^2}{2S_{fl}S_m} = \frac{(0.04)^2 + \left(\frac{2}{15}\right)^2}{2 \times 0.04 \times \frac{2}{15}} = 1.8167$$

(*c*) Let the external resistance per phase added to the rotor circuit be *r* ohms

∴ Rotor resistance per phase, $R_2(0.02 + r)$

The starting torque will be maximum when

$$R_2 = X_2$$

$0.02 + r = 0.15,\ r = 0.15 - 0.02 = 0.13\ \Omega$

Example 1.80

A 3-phase, 50 Hz, 6-pole induction motor runs at 940 rpm and delivers 7 kW output. What starting torque will the motor develop when switched directly on the supply, if maximum torque is developed at 800 rpm? The friction and windage laser total 840 W.

Solution:

$$S = \frac{N_s - N}{N_s} = \frac{1000 - 940}{1000} = 0.06$$

$$S_m = \frac{N_s - N_m}{N_s} = \frac{1000 - 800}{1000} = 0.2$$

$$P_{md} = 7000 + 840 = 7840 \text{ W}$$

$$P_{md} = 2\pi h \tau_d$$

$$7840 = 2\pi \times \frac{940}{60} \tau_{dfl}$$

$$\tau_{dfl} = \frac{7840 \times 60}{2\pi \times 940} = 79.645 \text{ Nm}$$

$$\tau_{st} = \frac{S^2 + S_m^2}{S(1 + S_m^2)} \tau_{dfl} = \frac{(0.06)^2 + (0.2)^2}{0.06(1 + 0.2)^2} \times 79.645 = 55.65 \text{ Nm}$$

Example 1.81

A 3-phase induction motor has a 4-pole, star-connected stator winding. The motor run on a 50 Hz. Supply with 200 V between lines. The rotor resistance and standstill rotor reactance per phase are 0.1 Ω and 0.9 Ω respectively. The ratio of rotor to stator turns is 0.67. Calculate

(a) total torque at 4% slip *(b) maximum torque*

(c) speed at maximum torque *(d) maximum mechanical power*

Neglect of stator impedance.

Solution: $E_2 = E_1\left(\frac{T_{e2}}{T_{e1}}\right) = \frac{200}{\sqrt{3}} \times 0.67 = 77.37 \text{ V}$

(*a*)

$$\tau_d = \frac{3sE_2^2R_2}{2\pi n_s[R_2^2 + (sX_2)^2]}$$

$$= \frac{3 \times 0.04 \times (77.37)^2 \times 0.1}{2\pi\left(\frac{1500}{60}\right)[(0.1)^2 + (0.04 \times 0.9)^2]} = 40.98 \text{ Nm}$$

(*b*)

$$\tau_{max} = \frac{kE_2^2}{2X_2} = \frac{3E_2^2}{2\pi n_s \times 2X_2}$$

$$= \frac{3 \times (77.37)^2}{2\pi \times \frac{1500}{60} \times 2 \times 0.9} = 63.51 \text{ Nm}$$

(*c*) Slip at maximum torque

$$S_m = \frac{R_2}{X_2} = \frac{0.1}{0.9} = \frac{1}{9}$$

Speed at maximum torque

$$N_m = (1 - S_m)N_s = \left(1 - \frac{1}{9}\right) \times 1500 = 1333.3 \text{ rpm}$$

(*d*)

$$(P_{md})_{max} = w_m\tau_{max} = \frac{2\pi N_m}{60}\tau_{max} = \frac{2\pi \times 1333.3}{60} \times 63.51$$

$$= 8867.5 \text{ Nm}$$

Example 1.82

A 3-phase induction motor has a starting torque of 100% and a maximum torque of 200% of the full-load torque. Determine

(a) Slip at which maximum torque occurs

(b) full-load slip

(c) rotor current at starting in per unit at full-load rotor current.

Solution: (*a*) $\dfrac{\tau_{st}}{\tau_{max}} = \dfrac{\tau_{fl}}{2\tau_{fl}} = \dfrac{1}{2}$

Also $\dfrac{\tau_{st}}{\tau_{max}} = \dfrac{2S_m}{1+S_m^2} = \dfrac{1}{2}$

$\therefore$ $S_m^2 - 4S_m + 1 = 0$

$$S_m = \frac{4 \pm \sqrt{16-4}}{2} = 3.732,\ 0.268$$

Neglecting the higher value

$$S_m = 0.268$$

(*b*)
$$\frac{\tau_{fl}}{\tau_{max}} = \frac{2s\,S_m}{S^2 + S_m^2}$$

$$\frac{1}{2} = \frac{2s\,S_m}{S^2 + S_m^2}$$

$$S^2 - 4s\,S_m + S_m^2 = 0$$

$$S = 4S_m \pm \sqrt{\frac{16S_m^2 - 4S_m^2}{2}}$$

Neglecting the higher value

$$S = 0.268\,S_m = 0.268 \times 0.268 = 0.0718$$

(*c*) Let the slip at full load be S

$$I_{2fl} = \frac{E_{2s}}{Z_{2s}} = \frac{SE_2}{\sqrt{R_2^2 + (SX_2)^2}}$$

At starting $S = 1$

$$I_{2st} = \frac{E_2}{\sqrt{R_2^2 + X_2^2}}$$

$\therefore$
$$\frac{I_{2st}}{I_{2fl}} = \frac{E_2}{\sqrt{R_2^2 + X_2^2}} \Bigg/ \frac{SE_2}{\sqrt{R_2^2 + (SX_2)^2}}$$

$$\left(\frac{I_{2st}}{I_{2fl}}\right)^2 = \frac{R_2^2 + (SX_2)^2}{S^2(R_2^2 + X_2^2)} = \frac{X_2^2\left(\dfrac{R_2^2}{X_2^2} + S^2\right)}{X_2^2\left(\dfrac{R_2^2}{X_2^2} + 1\right)S^2}$$

Since, $S_m = \dfrac{R_2}{X_2}$

$$\left(\frac{I_{2st}}{I_{2fl}}\right)^2 = \frac{S_m^2 + S^2}{S^2(1 + S_m^2)} = \frac{(0.268)^2 + (0.0718)^2}{(0.0718)^2(1 + 0.268^2)} = 13.9316$$

$\therefore$
$$\frac{I_{2st}}{I_{2fl}} = \sqrt{13.9316} = 3.7325$$

Example 1.83

A 3-phase induction motor with rotor resistance per phase equal to the standstill rotor reactance has a starting torque of 25 Nm. For negligible stator impedance and no-load current. Determine the starting torque in case the rotor circuit resistance per phase in (a) doubled (b) halved.

Solution:

$$R_2 = X_2$$

$$Z_{st} = \frac{kR_2}{R_2^2 + X_2^2}$$

$$25 = \frac{kR_2}{R_2^2 + R_2^2}$$

$$\therefore \quad k = 50\ R_2$$

(*a*) New rotor resistance $= 2R_2$

$$\therefore \quad \tau_{st} = \frac{k(2R_2)}{(2R_2)^2 + R_2^2} = \frac{50R_2 \times 2R_2}{5R_2^2} = 20 \text{ Nm}$$

(*b*) Rotor resistance $= \frac{1}{2} R_2$

$$\tau_{st} = \frac{k\left(\frac{1}{2}R_2\right)}{\left(\frac{1}{2}R_2\right)^2 + R_2^2} = \frac{50R_2 \times \frac{1}{2}R_2}{R_2^2 + R_2^2} = 20 \text{ Nm}$$

Example 1.84

The rotor resistance and reactance of a 4-pole, 50 Hz. 3-phase slip ring induction motor are 0.4 and 4 Ω/phase respectively. Calculate the speed at maximum torque and the ratio (man torque/(starting torque). What value should the resistance per phase have so that the starting torque is half of maximum torque?

Solution:

$$N_s = \frac{120f}{P} = \frac{120 \times 50}{4} = 1500 \text{ rpm}$$

$$S_m = \frac{R_2}{X_2} = \frac{0.4}{4} = 0.1$$

$$N_m = (1 - S_m)\ N_s = (1 - 0.2) \times 1500 = 1350 \text{ rpm}$$

$$\frac{\tau_{max}}{\tau_{st}} = \frac{1 + S_m^2}{2S_m} = \frac{1 + (0.1)^2}{2 \times 0.1} = 5.05$$

Let the additional resistance required per phase be $r\ \Omega$

$$\tau_{st} = \frac{k(R_2 + r)}{(R_2 + r)^2 + X_2^2}$$

$$\tau_{max} = \frac{k}{2X_2}$$

$$\frac{\tau_{st}}{\tau_{max}} = \frac{1}{2} = \frac{k(R_2 + r)}{(R_2 + r)^2 + X_2^2} \Big/ \frac{k}{2X_2} = \frac{1}{2}$$

$$= \frac{2X_2(R_2 + r)}{(R_2 + r)^2 + (X_2)^2} = \frac{1}{2}$$

$$= 2 \times 2 \times 4(0.4 + r) = (0.4 + r)^2 + (4)^2$$

$$6.4 + 16r = 0.16 + 0.8r + r^2 + 16$$

$$r^2 - 15.2r + 9.76 = 0$$

$$r = \frac{15.2 \pm \sqrt{(15.2)^2 - 4 \times 9.76}}{2}$$

$$= \frac{1}{2}(15.2 \pm 13.856) = 14.52 \text{ or } 0.807$$

The lower value r is taken otherwise R would be greater thus X_2 and the maximum torque would never be reached. For example, it $r = 14.52$, then for maximum torque.

$$14.52 = sX_2$$

$$S = \frac{14.52}{4} > 1$$

which is not possible

$$r = 0.807\ \Omega$$

Example 1.85

A 440 V, 50 Hz. Squirrel cage induction motor has a ratio of standstill reactance to resistance of rotor per phase of 3 to 2 and a maximum torque which is 4 times the normal full load torque. Calculate

(a) full load slip.

(b) ratio of starting torque of full-load torque.

(c) minimum voltage required to develop the normal full load torque at starting.

Solution:

$$S_m = \frac{R_2}{X_2} = \frac{1}{3}$$

$$\frac{\tau_{max}}{\tau_{fl}} = \frac{S_m^2 + S^2}{2s\,S_m}$$

$$4 = \frac{\left(\frac{1}{3}\right)^2 + S^2}{2S \times \frac{1}{3}}$$

$$4 = \frac{1 + 9S^2}{6S}$$

$$9S^2 - 24S + 1 = 0$$

$$S = \frac{24 \pm \sqrt{24^2 - 4\times9\times1}}{2\times9} = 0.0423 \text{ or } 2.624$$

$\therefore$ $$S = 0.0423$$

(*b*) $$\frac{\tau_{st}}{\tau_{fl}} = \frac{S^2 + S_m^2}{S(1+S_m^2)} = \frac{(0.0423)^2 + \left(\frac{1}{3}\right)^2}{0.0423\left(1+\frac{1}{9}\right)} = 2.402$$

(*c*) $\tau_{st} = \tau_{fl}$

$$\frac{kV_{11}^2 R_2}{R_2^2 + X_2^2} = \frac{RV_1^2\, sR_2}{R_2^2 + (sX_2)^2}$$

$$V_{11}^2 = \frac{SV_1^2(R_2^2 + X_2^2)}{R_2^2 + (SX_2)^2}$$

$$= \frac{0.04\left(\frac{440}{\sqrt{3}}\right)^2 (R_2^2 + 9R_2^2)}{R_2^2 + (0.04\times9R_2)^2} = \frac{0.04\times(440)^2\ (1+9)}{3(1+0.1276)}$$

$$V_{11} = 151.17 \text{ V}$$

$$\text{Line voltage} = \sqrt{3}\times151.17 = 261.8 \text{ V}$$

❑❑❑

Chapter

2 Single Phase Induction Motors

2.1 SINGLE PHASE MOTORS

As the name suggest these motors are used on single phase supply. Single phase motors are most familiar of all electric motors because they are extensively used in home appliances, shops, offices etc. It is true that single phase motor are less efficient substitute for 3ϕ motor but 3ϕ power is normally not available except in large commercial and industrial establishment. Every where 3ϕ mains are present single phase supply may be obtained by using one of three lines and the neutral. Domestic appliances like fans, hair driers, washing machine, vacuum cleaner, mixer, refrigerator, food processor and other kitchen equipment employ these motors.

Single phase motor are classified as follows:

1. Induction motor
2. Commutator motor
3. Synchronous motor

2.2 PRODUCTION OF ROTATING FIELD

Consider two winding *A* and *B* so displaced that they produce magnetic field 90° apart in space as shown in Fig. 2.1(*a*). Suppose that these winding produce magnetic field equal in magnitude and 90° apart in time given by

$$\phi_A = \phi_m \cos \omega t$$
$$\phi_B = \phi_m \sin (\omega t + 90°)$$

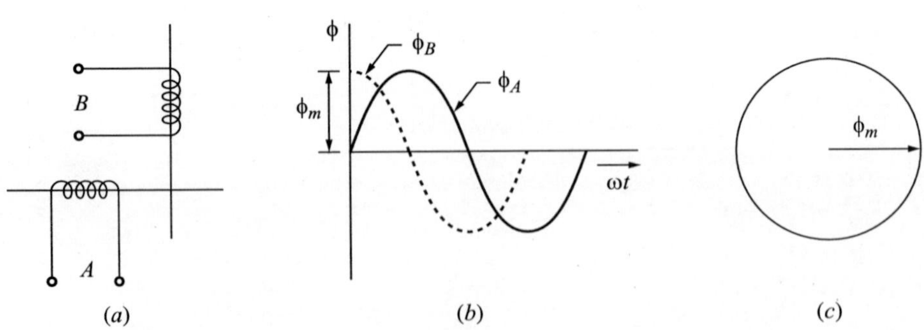

Fig. 2.1

The waveform of these fields are shown in Fig. 2.1(*b*). The resultant of these two field is a rotating magnetic field of constant magnitude ϕ_m.

Suppose that two windings *A* and *B* are displaced 90° in space but produce field that are either not equal or not 90° apart in time. The resultant of these field is again a rotating field but this field is variable in magnitude. A non-uniform magnetic field if produces a non-uniform torque which makes operation of motor noise. Also non-uniform rotating field has smaller starting torque as compared to uniform rotating field.

2.2.1 Single Phase Induction Motor Principle

A single phase induction motor consist of a single phase winding mounted on stator and a cage winding on the rotor. When a 1ϕ supply is connected to stator winding a stationary pulsating magnetic field is produced. By pulsating field we mean that fields builds up in one direction falls to zero and then builds up in the opposite direction under these condition the rotor does not rotate due to inertia. Therefore a single phase induction motor is inherently and not self starting and requires some special starting means. If however the single phase stator winding is eruted and the rotor of motor is started by an auxiliary means and the starting device is then removed, the motor continues to rotate in direction in which it is started.

Two theories have been suggested to analyse the performance of a single phase induction motor namely the double revolving field theory and the crust field theory. Both the theories are fairly complicated and neither has any advantage over the other in numerical calculation.

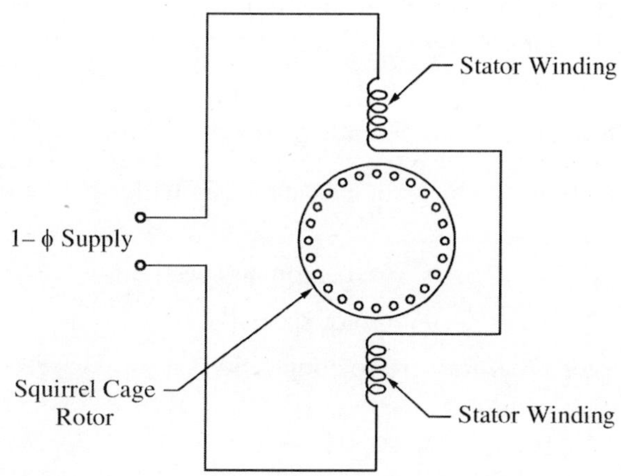

Fig. 2.2

2.2.2 Double Revolving Field Theory of Single Phase Induction Motor

According to double revolving field theory of single phase induction motors, stationary pulsating magnetic field can be resolved into two components of equal magnitude but rotating in opposite direction.

The equation for an alternating magnetic field axis is fixed in space is given by

$$k(\alpha) = \beta_{max} \sin \omega t \cos \alpha$$

β_{max} is maximum value of sinusoidally distributed flux density produced by stator winding carrying an alternating current of frequency ω and α is space displacement angle

$$\sin A \cos B = \frac{1}{2}\sin (A - B) + \frac{1}{2}\sin (A + B)$$

can be written as

$$k(\alpha) = \frac{1}{2}\beta_{max} \sin (\omega t - \alpha) + \frac{1}{2}\beta_{max} \sin (\omega t + \alpha)$$

First term represent revolving field in +ve α direction whereas IInd term represent revolving field is –ve α direction. Field and second field has maximum value $\frac{\beta_{max}}{2}$.

First field is called forward rotating field and second field is called backward rotating field. Both the field is rotating with synchronous speed in opposite direction.

At stationary position if rotor induced voltage are equal and opposite. Consequently the two torque are equal and opposite. Hence at standstill net require is zero. Therefore single phase induction motor is not self starting.

However if rotor is given an initial rotation by auxiliary means in either direction the torque due to rotating field acting in direction of initial rotation will be more hence motor will develop a net positive torque and start rotating in that direction.

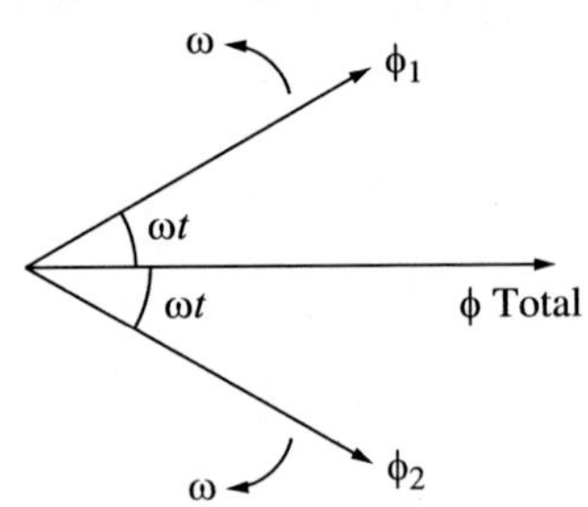

Fig. 2.3

2.2.3 Rotor Slip w.r.t. Two Relating Field

By defination direction in which rotor is started initially will be called the forward field.

Let n_s = synchronous speed

n = rotor speed

The slip of rotor w.r.t. forward rotating field is

$$s_f = s = \frac{n_s - n}{n_s} = 1 - \frac{n}{n_s} \quad ...(i)$$

For backward rotating flux sign if n must be charged

$$s_b = \frac{n_s - (-n)}{n_s} = \frac{n_s + n}{n_s} = 1 + \frac{n}{n_s} \quad ...(ii)$$

Adding Eqn. (*i*) and (*ii*)

$$s + s_b = 2$$

$$s_b = 2 - s$$

The rotor equivalent circuit for forward and backward rotating flux is shown in Fig. 2.4.

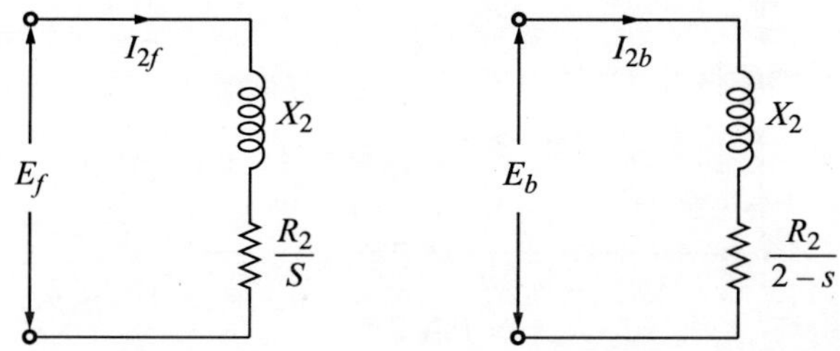

Fig. 2.4 Rotor equivalent circuit (a) for forward rotating flux (b) for backward rotating flux

- At standstill impedances are equal however when circuit rotates impedance of motor circuit are unequal and rotor current I_{2b} is greater than other current E_{2f}.
- Consequently as speed increases the forward flux increases while the backward flux decreases. However resultant flux remains essentially constant. This resultant flux induce voltage in stator winding. Both flux wave induce voltage in rotor and produce torque in rotor. These two torque are in opposite direction. The net induced torque in motor is equal to difference between these torque.

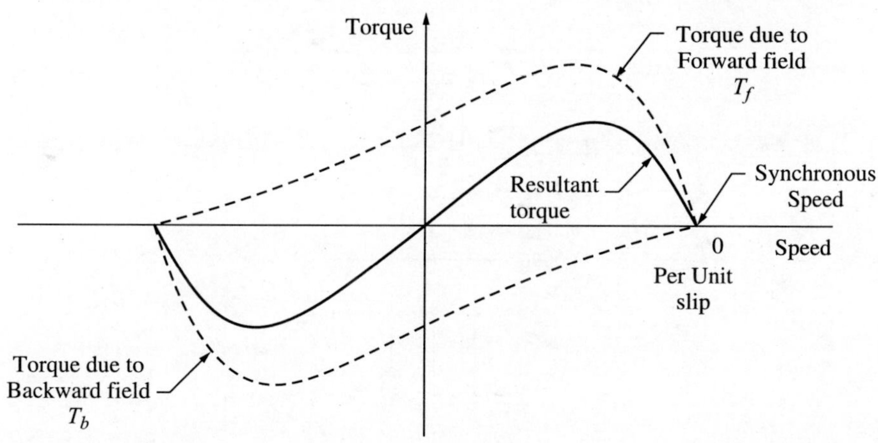

Fig. 2.5 Torque speed characteristic

2.2.4 Equivalent Circuit of Single Phase Induction Motor

Equivalent circuit of a 1ϕ induction motor with only its main winding energised at standstill.

R_{1m} = resistance of main stator winding

X_{1m} = leakage reactance of main stator winding

X_M = magnetizing reactance

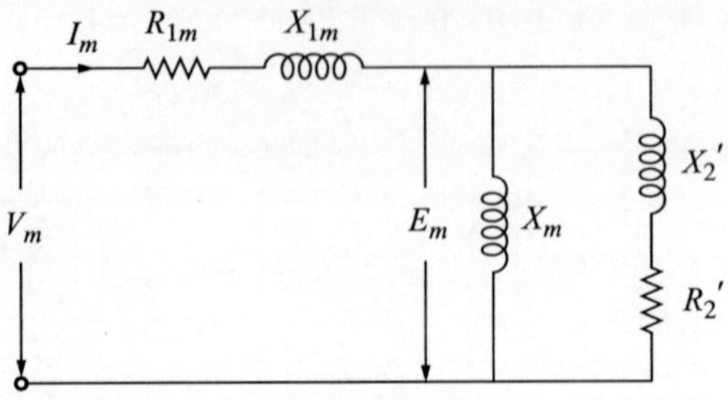

Fig. 2.6

R_2' = standstill rotor resistance referred to main stator winding.
X_2' = standstill rotor leakage reactance referred to main stator winding.
V_m = applied voltage
I_m = main winding current

According to double revolving

According to double revolving field theory the pulsating air gap flux in motor at standstill can be resolved into two equal and opposite flux with the motor. Thus a single ϕ induction motor may be considered as consisting of two rotor having a common stator winding.

$$2f = R_f + j\,X_f = \left(\frac{R_2'}{2s} + \frac{j\,X_2'}{2}\right) \Bigg\| \left(\frac{i\,X_m}{2}\right)$$

$$2b = R_b + j X_b = \left[\frac{R_2'}{2(2-s)} + \frac{j\,X_2'}{2}\right] \Bigg\| \left(\frac{j\,X_m}{2}\right)$$

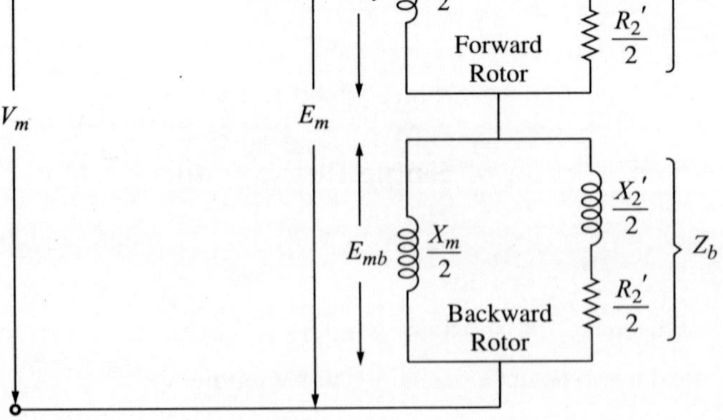

At standstill with the effects of forward and backward rotating fluxer separated

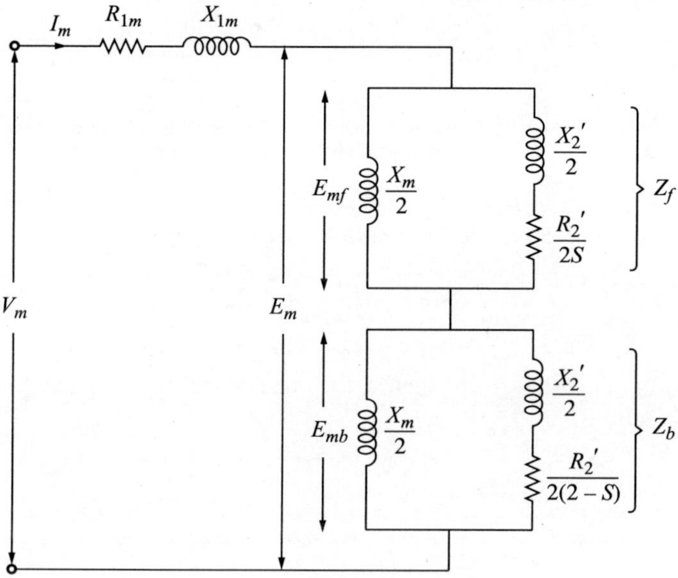

During Running Conduction

Fig. 2.7

2.2.5 Performance Calculation of a Single Phase Single Winding Induction Motor

The equivalent circuit is similar to 3φ induction motor with the exception that there are both forward and backward components of power and torque present in it. The relationship for power and torque obtained for 3φ induction motor can also be applied for either the forward or backward components of 1-φ induction motor.

Total air gap power

$$Pg = Pg_f - Pg_b$$

$$\because \quad Pg_f = E_m^{\ 2} R_f$$

$$Pg_b = E_m^2 R_b$$

$$Pg = I_m^{\ 2} R_f - I_m^{\ 2} R_b = I_m^2 \ (R_f - R_b)$$

The torque produced by forward field

$$\tau_f = \frac{P_{gf}}{w_s} = \frac{P_{gf}}{2\pi n_s}$$

The torque produced due to backward field

$$\tau_b = \frac{P_{gb}}{w_s} = \frac{P_{gb}}{2\pi n_s}$$

where w_s = synchronous speed in rad/s.

Resultant electromagnetic torque

$$\tau_{\text{ind}} = \tau_f - \tau_s$$

Now $\tau_{\text{ind}} = \dfrac{P_g}{w_s} = \dfrac{P_{gf} - P_{gb}}{w_s} = \dfrac{I_m^2(R_f - R_b)}{w_s}$

Total as loss is the sum of rotor as loss due to forward field and rotor as loss due to backward field.

$$P_{rc} = P_{rcf} + P_{rcb}$$

Rotor as loss in a 3ϕ induction motor = slip × air gap power

$$P_{\text{ref}} = sP_{gf}$$
$$P_{rcb} = (2 - s)\, P_{gb}$$
$$P_{rc} = s\, P_{gf} + (2 - s)\, P_{gb}$$

The power converted from electrical to mechanical form in a single ϕ induction motor is given by

$$P_{\text{mech}} = P_{\text{conv}} = w\, \tau_{\text{ind}}$$

$\because$ $w = (1 - s)\, w_s$

$$P_{\text{mech}} = P_{\text{conv.}} = (1 - s)\, w_s\, \tau_{\text{ind}}$$
$$= (1 - s)\, P_g = (1 - s)\, (P_{gf} - P_{gb})$$
$$P_{\text{mech}} = I_m^2(R_f - R_b)\, (1 - s)$$
$$P_{\text{out}} = P_{\text{conv.}} - \text{core loss} - \text{mechanical loss} - \text{stray losses}$$
$$P_{\text{out}} = P_{\text{mech}} - P_{\text{rot}}$$
$$P_{\text{rot}} = \text{rotational loss} = \text{friction loss} + \text{windage loss} + \text{core loss}$$

2.3 DETERMINATION OF EQUIVALENT PARAMETERS

We can find parameters of 1-ϕ induction motor using blocked rotor and no load test. These tests are similar to those made in 3ϕ induction motor. Except far capacitor run motor these test are performed with auxiliary winding kept open.

2.3.1 Blocked Rotor Test

Blocked motor means rotor is at rest. A low voltage is applied to stator so that rated current flows in main winding. We will measure the voltage current and power respectively under those condition. Let V_{sc}, I_{sc} and P_{sc} denote the voltage, current and power respectively under these conduction with rotor blocked, $s = 1$ the impedance $\dfrac{X_m}{2}$ in the equivalent circuit is so large compared with $\left(\dfrac{R_2'}{2} + \dfrac{jX_2'}{2}\right)$ that it may be neglected from equivalent circuit.

The equivalent impedance referred to stator is given by

$$\tau_c = \frac{V_{sc}}{I_{sc}}$$

The equivalent series resistance R_c if motor is

$$R_c = R_{1m} + \frac{R_2'}{2} + \frac{R_2'}{2} = R_{1m} + R_2' = \frac{P_{sc}}{I_{sc}^2}$$

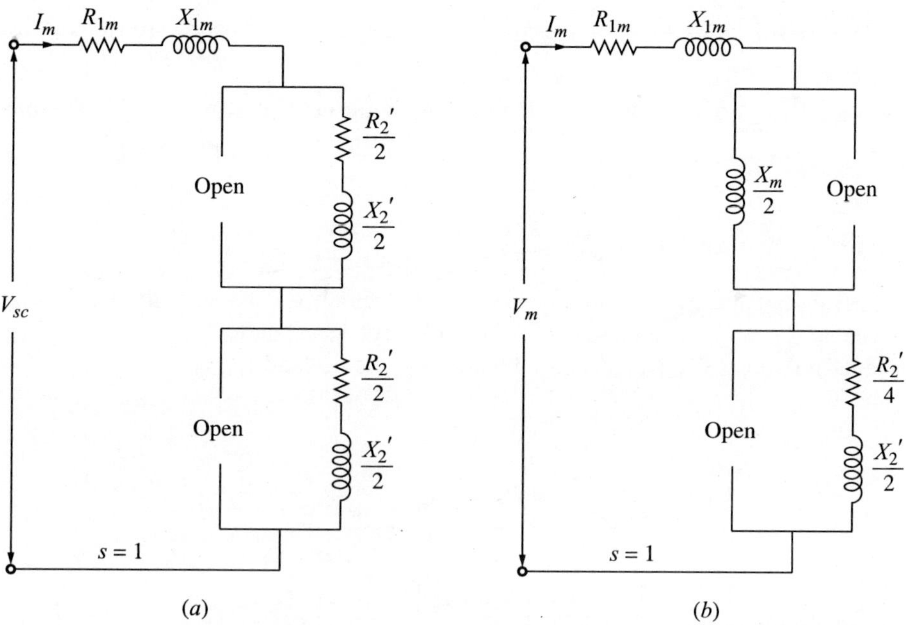

Fig. 2.8 Simplified equivalent circuit of a 1ϕ induction motor (*a*) with rotor locked (*b*) at no load

∵ Resistance of main stator winding R_{1m} is known to us the effective rotor resistance will be

$$R_2' = R_e - R_{1m} = \frac{P_{sc}}{I_{sc}^2} - R_{1m}$$

the equivalent reactance X_e is given by

$$X_e = X_{1m} + \frac{X_2'}{2} + \frac{X_2'}{2} = X_{1m} + X_2'$$

Since leakage reactance X_{1m} and X_2' cannot be separated. We assume that $X_{1m} = X_2'$

$$X_{1m} = X_2' = \frac{1}{2} X_e = \frac{1}{2}$$

Thus from blocked rotor test parameters R_2', X_{1m}, X_2' can be found of R_{1m} is known.

2.3.2 No Load Test

The motor is run without load at rated voltage and rated frequency. The voltage, current and output power are measured. At no load, slip is very small close to zero and $\frac{R_2'}{2s}$ is very large as compared to $\frac{X_M}{2}$.

The resistance $\frac{R_2'}{2(2-s)} = \frac{R_2'}{2(2-0)} = \frac{R_2'}{4}$ associated with backward rotating field is so small as compared to $\frac{X_M}{2}$ that magnetizing current is negligible. Therefore under no load conduction equivalent circuit will become as in Fig. 2.8(*b*) equivalent reactance at no load

$$X_0 = X_{1M} + \frac{X_M}{2} + \frac{X_2'}{2} \qquad ...(1)$$

Since X_{1m} and X_2' are already known from blocked rotor test, the magnetizing reactance X_M can be calculated from Eqn. (1). Let V_0, I_0 and P_0 denote the voltages current and power respectively in no load test. The no load p.f. is

$$\cos \phi_0 = \frac{P_0}{V_0 I_0}$$

The no load equivalent impedance is

$$Z_0 = \frac{V_0}{I_0}$$

The no load equivalent reactance is

$$X_0 = Z_0 \cos \phi_0 = Z_0 \times \sqrt{1 - \cos^2 \phi_0}$$

2.4 STARTING METHODS AND TYPES OF SINGLE PHASE THAN MOTOR

We have seen that some means should be used to start the single phase induction motor. Mechanical methods are unpractical and therefore motor is started temporarily converting it into a two phase motor single phase induction motor are usually classified according to auxiliary means used to start the motor. They are classified as follows:

1. Split phase motor
2. Capacitor start motor
3. Capacitor start – capacitor run motor
4. Permanent split capacitor motor
5. Shaded pole motor.

All these starting methods depend on two alternating fields displaced in space and phase.

The resultant of two field is a rotating field. This rotating field reacts with the cage rotor to provide starting torque, one field is produced by the main winding and the other by auxiliary winding. The auxiliary winding is also called starting winding.

2.4.1 Split Phase Induction Motor

Figure 2.9 shows a split phase induction motor. It is also called a resistance start motor. It has a single cage rotor and its stator has two windings—a main winding and a starting winding. The main field winding and the starting winding are displaced 90° in space like the windings in a two phase induction motor. The main winding has a very low resistance and high inductive reactance.

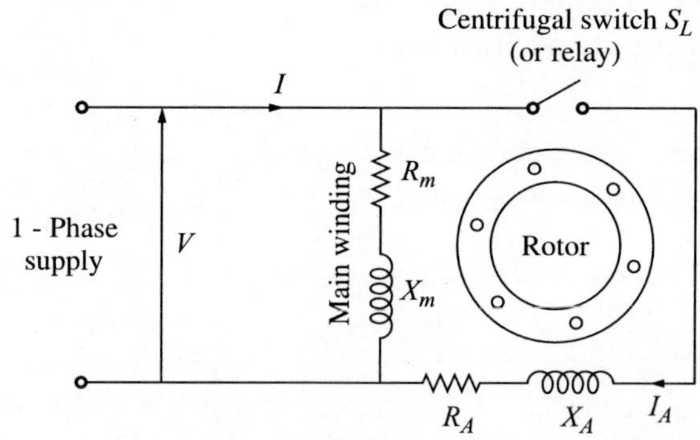

Fig. 2.9

Thus the current I_M is main winding lags behind the supply voltage V by nearly 90°. The auxiliary winding has a resistor connected in series with it. It has a high resistance and low inductive reactance so that current I_A is auxiliary winding is nearly in phase with line voltage. Thus there is time phase difference between the current in the two windings. The time phase difference ϕ is not 90° but usually of the order of 30°. This phase difference is enough to produce a rotating magnetic field. Since the current in the two windings are not equal, the rotating field is not uniform and starting torque is small of the order of 1.5 to 2 times the rated running torque. The main and auxiliary winding are connected in parallel during starting. The starting winding is disconnected from the supply automatically when the motor reaches speed about 70 to 80% of synchronous speed. For motors about 100W or more a centrifugally operated switch is used to disconnect the starting winding. For smaller motor a relay is often used. At the time of starting, a heavy current flows in the relay coil causing its contacts to close. This bring the starting winding into the circuit. As the motor reaches its predetermined speed of the order of 70 to 80 percent of synchronous speed, the current through relay coil decreases consequently the relay opens and disconnect the auxiliary winding from the main supply and the motor then runs only on main winding. The torque speed characteristic of this motor is shown is Fig. 2.10(a). Figure 2.10(b) shows speed n_0 at which the centrifugal operates.

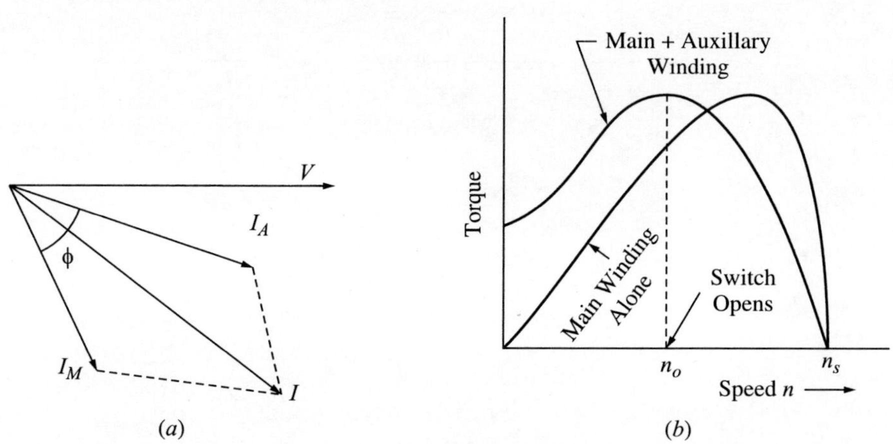

Fig. 2.10

Reversal of Direction of Rotation

This motor continues to run in the direction in which it is started. The direction of rotation of resistance start induction motor may be reversed by reversing the line connection of either the main winding on the starting winding. The motor must be bought to rect for this purpose. That is reversal of rotation can be made only when the motor is standstill but not while running.

Application

Split phase motors are cheap. These motors are suitable where a moderate starting torque is required *e.g.*, (*a*) fans (*b*) washing machine (*c*) oil burneys (*d*) small machine tools, etc. The power rating of such motor lies between 60 W and 250 W.

2.4.2 Capacitor Motors

Capacitor motors are single phase induction motors that employ a capacitor in the auxiliary winding circuit to produce a greater phase difference between the current in the main and auxiliary winding. There are three types of capacitors motors.

2.4.2.1 Capacitor Start Motors

Figure 2.11 shows the connection of a capacitor start motor. It has a cage rotor and its stator has two winding namely, the main winding and the auxiliary winding (starting winding). The two winding are displaced 90° in space. A capacitor C_S is connected in series with starting winding. A centrifugal switch S_C is also connected as shown.

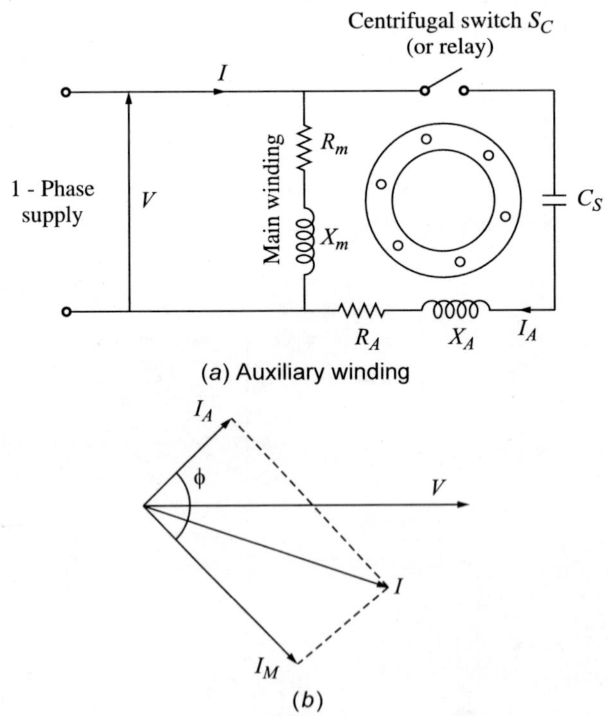

Fig. 2.11

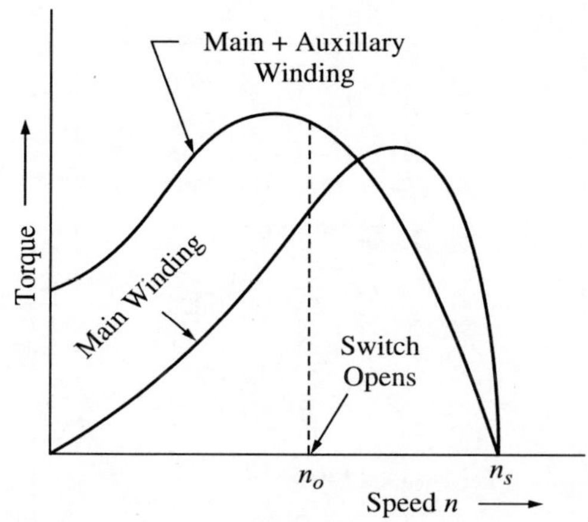

Fig. 2.12

By choosing a capacitor of proper rating the current I_M in the main winding may be made to lag the current I_A in the auxiliary winding by 90°. Thus the winding are displaced 90° electrical and their mmf are equal in magnitude but 90° apart in time phase. The motor is so named because it uses the capacitor only for the purpose of starting.

Motor Characteristic

The capacitor start motor develops a much higher starting torque (3.0 to 4.5 times full load torque) than does an equally rated resistance start motor. Capacitor start motors are more costly than split phase motor because of additional cost of capacitor. The torque speed characteristic of motor capacitor shown in Fig. 2.11(*b*).

Reversal of Direction of Rotation

The capacitor start motor may be reversed by reversing the connection if one of the winding. The motor is first brought to rect for this purpose.

Application

Capacitor start motor are used where high starting torque is required and where starting period may be long.

(*a*) compressor

(*b*) large fans

(*c*) pumps

(*d*) high inertia loads.

2.4.2.2 Two Value Capacitor Motors

Figure 2.13 shows diagram of a two valve capacitor motor. It has a cage rotor and its stator has two winding viz. main winding and the auxiliary winding (starting winding). The two winding are displaced 90° in space. The motor uses two capacitors C_S and C_R. The two capacitor are connected in parallel at starting.

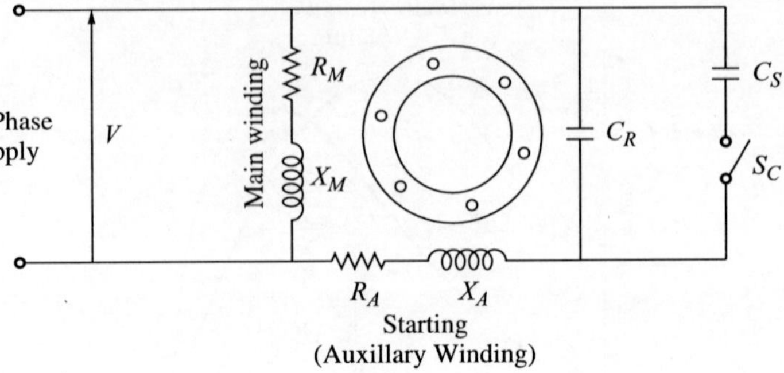

Fig. 2.13

The capacitor C_S is called starting capacitor. In order to obtain high starting torque, a large current is required for this purpose, the capacitive reactance X in the starting winding should be low. Since $X_A = \dfrac{1}{2\pi f C_A}$ the value of C_S should be large. The capacitor C_S is converted in circuit for short time during starting.

During normal operation current is smaller as compared to starting. Hence value of capacitance is small. As the motor reaches synchronous speed capacitor C_S is disconnected by a centrifugal switch S_C. The capacitor C_R is permanently connected in the circuit. It is called run capacitor. It is long time rated for continuous running. Hence this motor is called capacitor start-capacitor run motors.

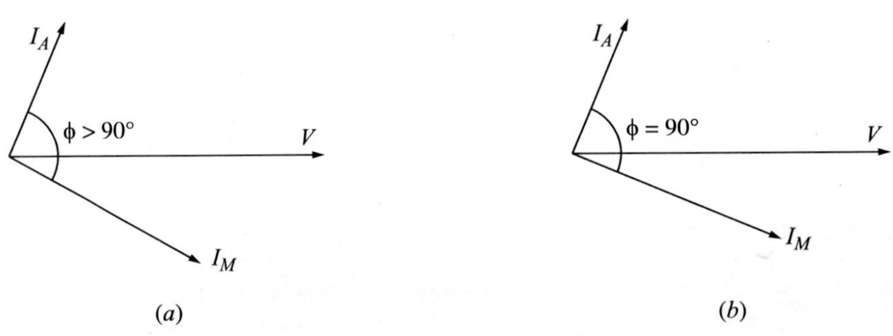

Fig. 2.14

Figure 2.14 (*a*) and (*b*) shows phasor diagram of a 2 value capacitor motor.

At starting both the capacitor are in circuit and $\phi > 90°$. When capacitor C_S is disconnected ϕ becomes 90° (electrical).

The torque speed curve is shown in Fig. 2.15.

Advantages

1. These motors are quiet and have smooth running.
2. They have a higher efficiency than motor that run as main winding above.

Application

Two value capacitors motors are used for loads of higher inertia requiring frequent chart where maximum pull out torque and efficiency required are higher. They are used at places where silence is important *e.g.*, hospitals, studies, etc.

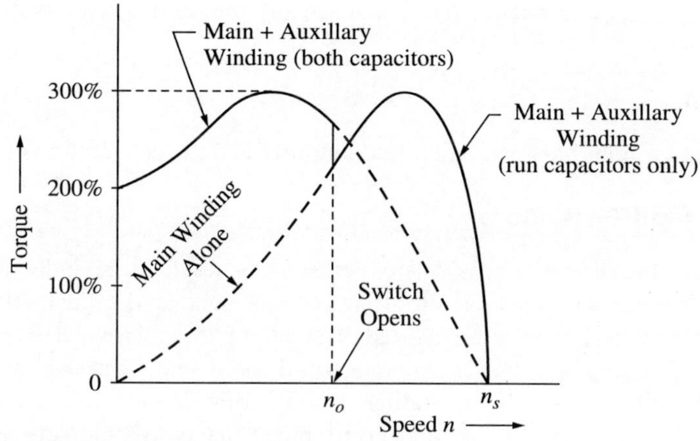

Fig. 2.15 Torque speed characteristic of a 2 value capacitor motors

2.4.3 Permanent Split Capacitor Motor

It has a cage rotor and its stator has two windings namely main winding and auxiliary winding. In this there capacitor which is permanently connected with auxiliary winding (*i.e.*, both starting and running condition). A permanent split capacitor motor is also called single value capacitor motor. This type of motor has no starting switch. The auxiliary winding is always in the circuit and therefore this motor operates in the same way as a balanced two phase motor. Therefore it is less noisy and produces a uniform torque.

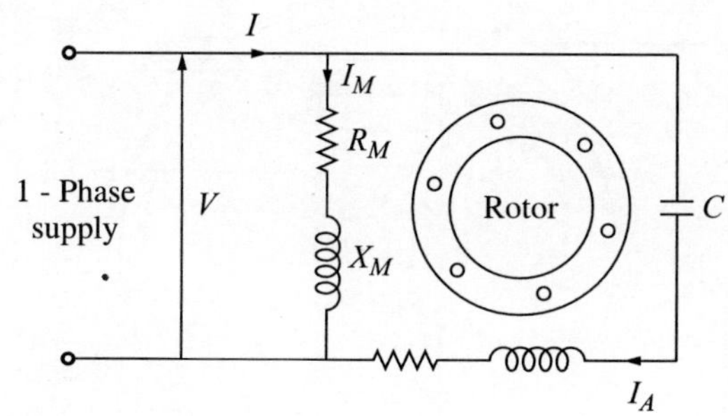

Fig. 2.16

Advantages

(*i*) No centrifugal switch is required.
(*ii*) It has higher efficiency.
(*iii*) It has higher power factor because of permanently connected capacitor.
(*iv*) It has a higher pull out torque.

Limitation

(*i*) Electrolytic capacitors cannot be used for continuous running. Therefore paper spaced VIR. filled type capacitors are to be used. Paper capacitors of equivalent rating are larger in size and more costly.

(*ii*) A single value capacitor has a low starting torque usually less than full load torque.

Application

They are used in fan, blowers, and air-conditioner and in refrigeration compressor.

2.4.4 Shaded Pole Motor

A shaded pole motor is a sample type of self starting single phase induction motor. It consist of a stator and a cage type rotor. The stator is made up if salient pole. Each pole is slatted on side and a curve ring is fitted on smaller part as shown. This part is called shaded pole. The ring is usually a single turn coil and is known as shading coil. When alternating current flows in field winding an alternating flux is produced in field core. A portion of this flux links with shading coil, which behaves as a short circuit secondary of a transformer. A voltage is induced in shading coil and this voltage circulates a current in it. The induced current produces a flux called induced flux which opposes the main core flux. The shading coil thus causes the flux in shaded portion *a* to lag behind the flux in unshaded portion *b* if the pole. At the same time the main flux and shaded pole flux are displaced in space. The space displacement is less than 90°. Since where is time and space displacement between the two flux, the condition for setting a rotating magnetic field is produced. Under the action if a rotating flux a starting torque is developed on the cage rotor. The direction of this rotating flux is from unshaded to the shaded portion of the pole. In Fig. 2.17 direction of rotation is clockwise. *In a shaded pole motor the reversal of direction if rotation is not possible.*

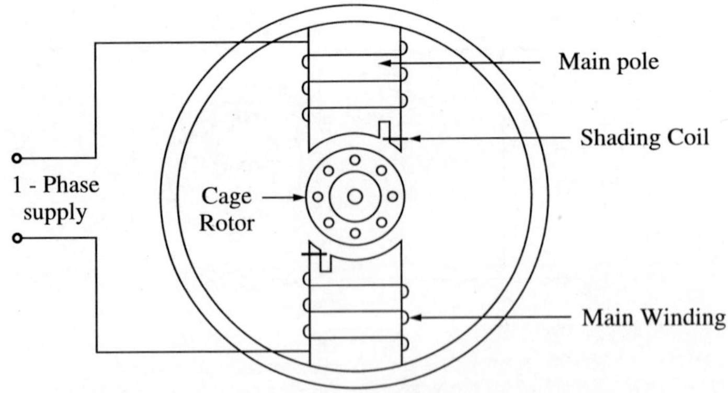

Fig. 2.17 Shaded pole motor with two stator poles

Application

They are built only in small sizes of power rating of the order if 40*W*, or less. They are used to drive device which have low starting torque the most common application are exhaust fans, table fans, hair driers, fan, for refrigeration and air conditioning equipment, cooling fans etc. They are also used in record players, tape recorder, photocopying machine, etc.

Advantages

(*i*) They are very cheap.

(*ii*) Extremely simple construction and absence of centrifugal switch.

Disadvantages

(*i*) Starting torque is very low.

(*ii*) Losses are high.

(*iii*) Low power factor.

2.5 COMPARISON BETWEEN 1-φ AND 3-φ INDUCTION MOTOR

Advantages of 3-φ Motor

(*i*) 1-φ motor develop about 50% of the output of that of a 3-φ motors for the same size and temperature rise.

(*ii*) 1-φ motor have low power factor.

(*iii*) The starting torque is low in 1-φ induction motor.

(*iv*) 1-φ motor have lower efficiency.

(*v*) 1-φ motor are costlier than 3-φ motors of the same rating.

Advantages of 1-φ Induction Motor

(*i*) They have simple and robust construction.

(*ii*) They are less expensive for smaller rating.

(*iii*) They are used in low power drives in small industries and domestic and commercial application. They are generally available upto 1 kW rating.

2.6 SINGLE PHASE SERIES (UNIVERSAL) MOTOR

- A dc series motor coil rotate in same direction even if we reverse the polarity of the line terminals because in series motor both armature and field winding are in series. Since armature current and flux reverse simultaneously torque always acts in same direction.
- It might be expected that dc series motor would operate on alternating current also. The direction of torque developed is a dc series motor is determined by both field polarity and direction of current through the armature ($T \propto \phi\, i_a$).
- The nature of torque developed in dc series motor will be pulsating and frequency will be twice the line frequency.
- Thus a series motor can run both on dc and ac. Motor that can be used with a single phase ac source as well as dc source of supply voltages are called universal motor.
- However a series motor which is specifically designed for dc operation suffers from following drawbacks when it is used on 1-φ ac supply:

  (*i*) It's efficiency is low due to hysteresis and eddy current loss.

(*ii*) The power factor is low due to large reactance of the field and armature winding.

(*iii*) Sparking of brushes is expensive.

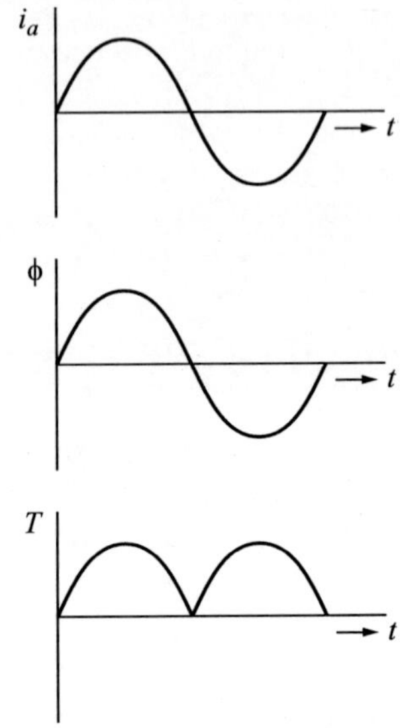

Fig. 2.18 Developed torque in a 1-ϕ series motor

In order to overcome these difficulties the following modification are made in dc series motor that is to operate satisfactorily in alternating current.

(*a*) The field core is constructed of material having low hysteresis loss. It is laminated to reduce eddy current loss.

(*b*) The field winding is provided with smaller number of turns. The field pole area is increased so that flux density is reduced. This reduces iron loss and reactive voltage drop.

(*c*) The number of armature conductor is increased in order to get required torque with low flux.

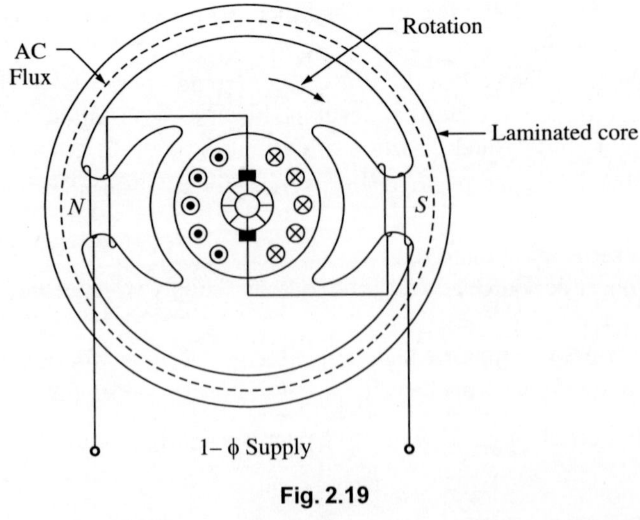

Fig. 2.19

(*d*) In order to reduce the effect of armature reaction thereby improving commutation and reducing armature reactance a compensating winding is used. This winding is put in stator starts as shown in Fig. 2.20.

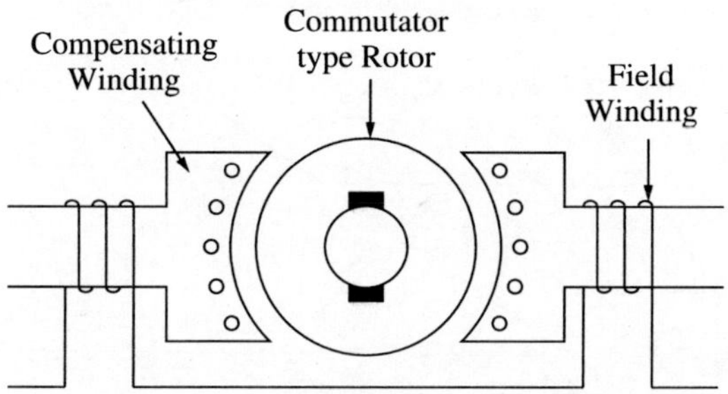

Fig. 2.20

- Compensating winding may be connected in series with both armature and field as shown. In such a core motor is conductively compensating.

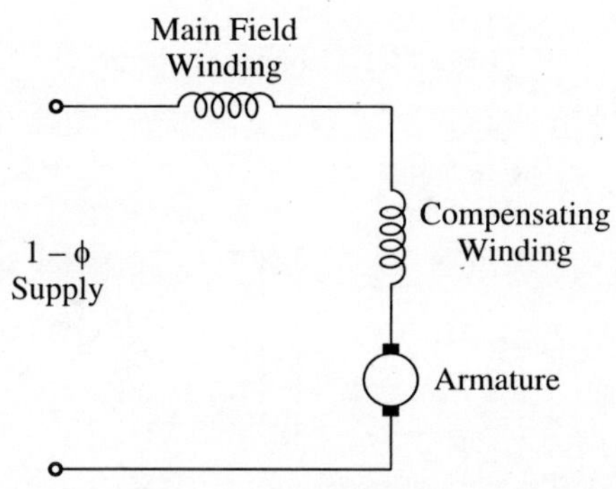

Fig. 2.21

- Compensating winding may be short circuited on itself in such core motor is said to be inductively compensated.
- The resistance drop $I_a R_{se}$, $I_a R_i$, $I_a R_e$ and $I_a R_a$ due to resistance of series field, interpole, compensating winding and armature respectively are in phase with armature current I_a.
- The reactance drop $I_a X_{se}$, $I_a X_i$, $E_a X_e$ and $I_a X_a$ due to reactance of series field, interpole winding, compensating winding and armature respectively lead current I_a by 90°. The generated armature counter emf is E_g. The terminal phase voltage V_p is equal to phasor sum of E_g and all impedance drop in series.

$$V_p = E_g + I_a Z_{se} + I_a Z_i + I_a Z_c + I_a Z_a$$

- In order to minimize commutation problem high resistance brush with increased brush area is used.

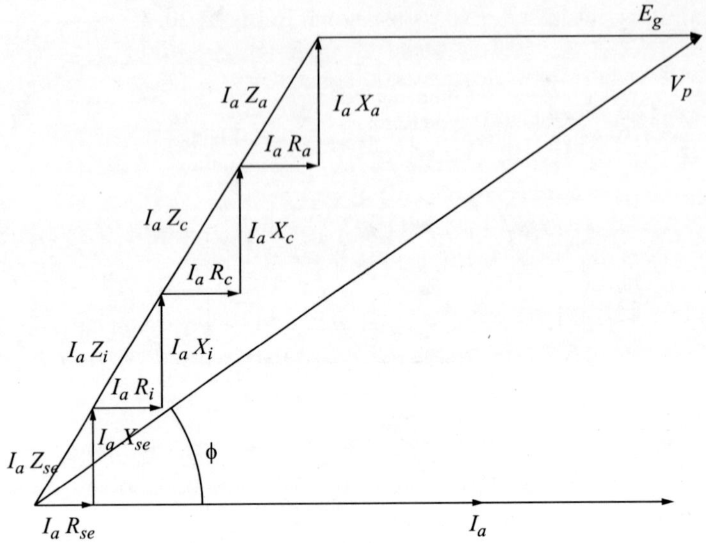

Fig. 2.22 Phasor diagram

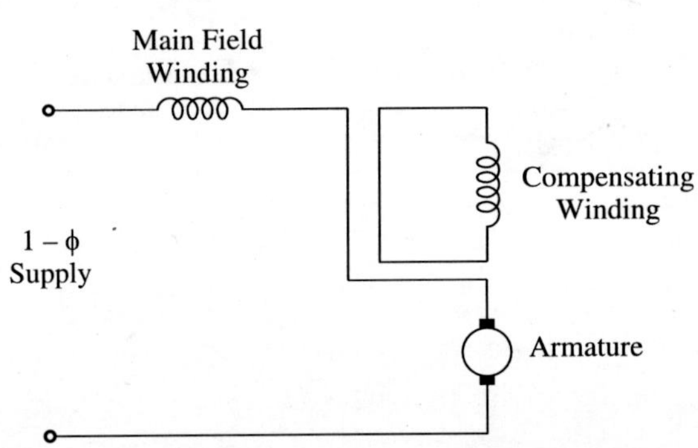

Fig. 2.23 Series motor with inductively compensated winding

Application

The fractional horse power ac series motor have high speed and large starting torque. They can therefore be used to drive:

(*a*) High speed vacuum cleaners
(*b*) Sewing machine
(*c*) Electric shavers
(*d*) Drills

Phasor diagram of AC Series Motor

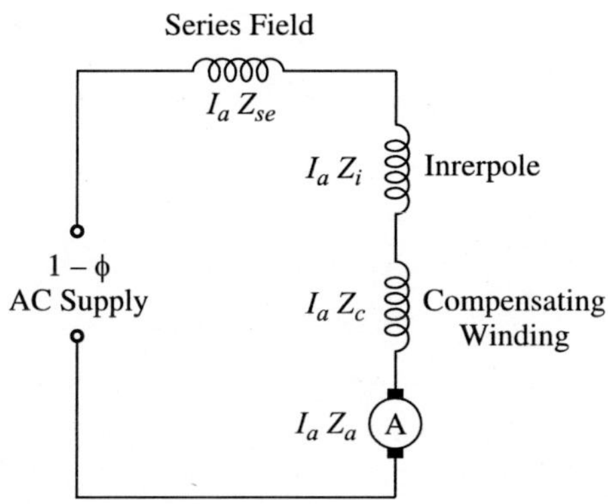

Fig. 2.24 Conductively coupled ac series motors

2.7 TWO PHASE AC SERVOMOTORS

Servomotors: Servomotors are used in feedback control systems. Servomotors have low rotor inertia and high speed of response. The servomotors are also known of control motors. The servomotors which are used in feedback control system should have linear relationship between electrical control signal and rotor speed, torque speed characteristic should be linear, the responsed the servomotors should be fast and inertia should be low.

Their power ratings vary from a fraction of a watt to a few hundred watt. The rotor of servomotors are designed with relatively long rotor length and smaller diameters. They generally operate at very low speeds and sometimes zero speed. They have larger size than that of conventional motors of similar power rating.

Types of Servomotors: The servomotors are classified as:

(*i*) AC servomotors
(*ii*) DC servomotors
(*iii*) Special servomotors

DC servomotors are further classified of armature controlled dc servomotors and field control dc servomotors.

DC Servomotors: DC servomotors are separately excited dc motors or permanent magnet dc motors. The speed of dc servomotors is normally controlled by varying the armature voltage. The armature of a dc servomotor has a large resistance so that the torque speed characteristic are linear and have a large negative slope (torque reducing with increasing speed). This provides a fast torque response because torque and flux become decoupled. Therefore, a step change in the armature voltage or current produces a quick change in the position or speed of the rotor.

The power rating of dc servomotors can vary from a few watts to several hundred watts. In general, high power servomotors are dc servomotors.

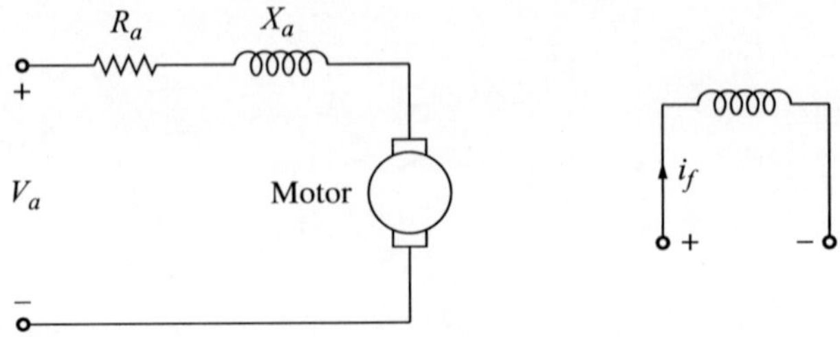

(*a*) Schematic diagram

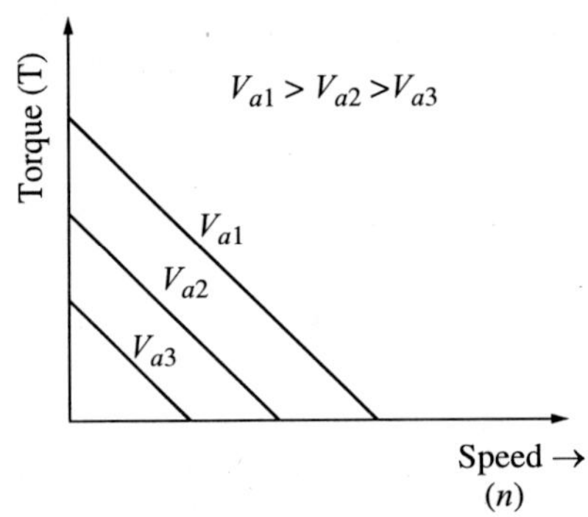

(*b*) Torque-speed characteristics

Fig. 2.25

2.7.1 AC Servomotors

These motors having two parts namely stator and rotor. AC servomotors are two phase induction motor. The stator has two distributed winding. These winding are displaced from each other by 90° electrical. One winding is called main winding or reference winding. The reference winding is excited by constant ac voltage, the other winding is called control winding. This winding is excited by variable control voltage of the same frequency as the reference winding but having a phase displacement of 90° electrical. The variable control voltage for control winding is obtained from a servo amplifier. The direction of rotation depends upon phase relationship of voltages applied to the two windings. The direction of rotation of rotor can be reversed by reversing the phase difference between control voltage and reference voltage.

The rotor of ac servomotors are of two types (*a*) squirrel cage rotors (*b*) drag cup type rotor. The squirrel cage rotors having large length and small diameter, so its resistance is very high. The air gap of squirrel cage is kept small. In drag cup type there are two air gap. For the rotor a cup it non magnetic material is used. A stationary iron core is placed between the conducting cup to complete the magnetic circuit. The resistance of drag cup type is high and therefore having high starting torque. Generally aluminium is used for cup.

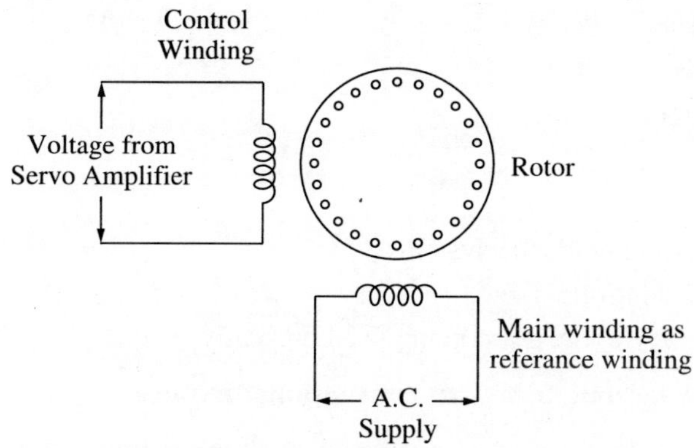

Fig. 2.26 AC servomotor

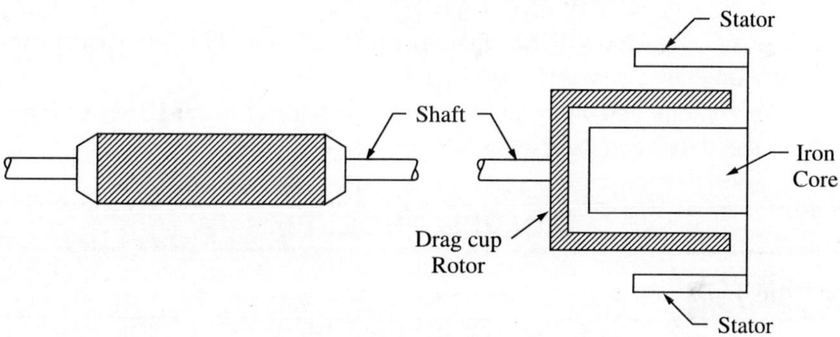

Fig. 2.27

Performance Characteristic of Two Phase AC Servomotor

The torque speed characteristic of two phase induction motor depends upon the ratio of reactance to resistance. For large $\frac{X}{R}$ characteristic is non-linear and for small $\frac{X}{R}$ it is linear. The torque speed characteristic for various control voltages are almost linear.

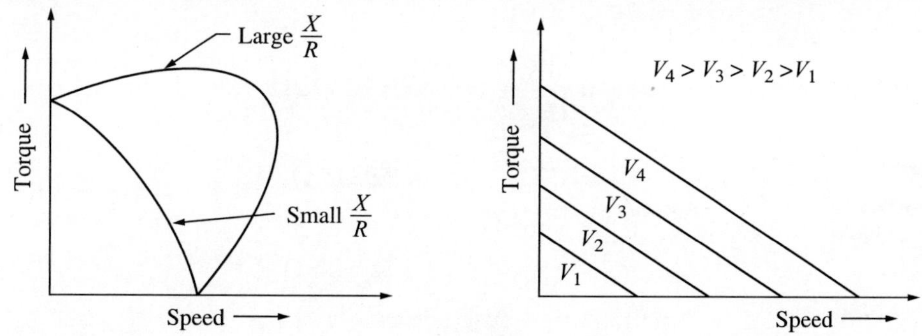

Fig. 2.28

Application of servomotors: These are following applications of servomotors:

(*i*) in rotors
(*ii*) computers
(*iii*) robots
(*iv*) machine tools
(*v*) training and guidance systems
(*vi*) process controllers, etc.

Both dc and ac servomotors are being used presently.

Comparison of servomotors with conventional motors

Servomotors have the following advantages over large industrial motors:

1. Servomotors produce high torques at all speeds including zero speed.
2. They accelerate or decelerate quickly.
3. They have low rotor inertia, therefore their direction of rotation can be reversed very quickly.
4. A servomotor can withstand higher temperature at lower shows or zero speed. As it can dissipate heat quickly.

SOLVED PROBLEMS

Example 2.1

A 220 A, 1 phase induction motor has following test results.

N.L. test V = 200 V, I = 6 A, P = 350 W

Locked rotor test: V = 125 V, I = 15 A, P = 580 W

Stator resistance: 1.2r with direct current.

Estimate the P.f. and efficiency when the slip is 0.05.

Solution: From locked rotor test

$$V_{SC} = 125 \text{ V}, \quad I_{SC} = 15 \text{ A}, \quad W_{SC} = 580 \ \Omega$$

$$Z_{eq} = \frac{V_{SC}}{I_{SC}} = \frac{125}{15} = 8.333 \ \Omega$$

and

$$R_{eq} = \frac{W_{SC}}{(I_{SC})^2} = \frac{580}{(15)^2} = 2.5778 \ \Omega$$

$$X_{eq} = \sqrt{Z_{eq}^2 - R_{eq}^2} = 7.9245 \ \Omega$$

Now R_1 = with Q.C.(... 1.5 times more)

$R_1 = 1.5 \times 1.2 = 1.8 \ \Omega$ with Q.C.

Now $R_{eq} = R_1 + R_2$

$R_2 = 2.5778 - 1.8 = 0.7778 \ \Omega$

$X_{eq} = X_1 + X_2 = 2X_1 = 2X_2$ (assume $X_1 = X_2$)

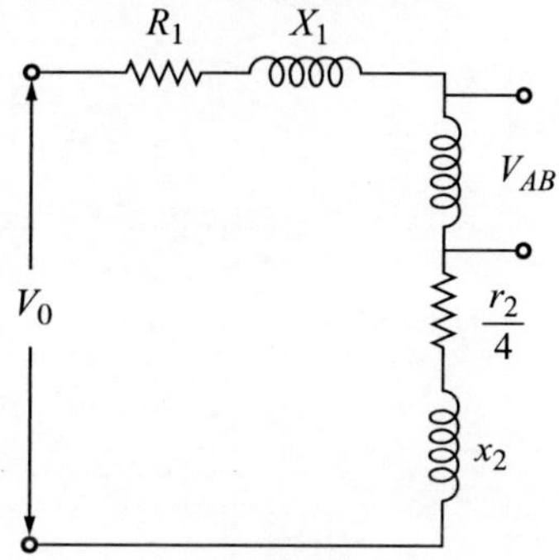

Fig. 2.29

$$X_1 = X_2 = \frac{X_{eq}}{2} = 3.9622\ \Omega$$

Now $$r_2 = \frac{R_2}{2} = 0.3889\ \Omega$$

$$X_2 = \frac{X_2}{2} = 1.3811\ \Omega$$

The equivalent circuit on no load is

$$I_0 = 6\text{ A}$$

$$\cos\phi_0 = \frac{w_0}{V_0 I_0} = \frac{350}{220\times 6} = 0.2651$$

$$\phi_0 = 74.62°$$

$$I_0 = 6\angle -74.62°\text{ A (Negative as lagging)}$$

$$V_0 = 220\angle 0°\text{ V}$$

$$\therefore \quad V_{AB} = \bar{V_0} - \bar{I_0} \times \left[\left(R_1 + \frac{r_2}{2}\right) + J(r_1 + r_2)\right]$$

$$= 220\angle 0° - 6\angle -74.62°\left[\left(1.8 + \frac{0.3889}{2}\right) + J(3.9622 + 1.9811)\right]$$

$$= 220\angle 0° - 6\angle -74.62° \times 6.269\angle 71.45°$$

$$= 220\angle 0° - 37.614\angle -3.17°$$

$$= 220 + J\,0 - 37.556 + J2.08 = 182.444 + J2.08\text{ V}$$

$$= 183.455\angle 0.653°\text{V}$$

$$V_{AB} = I_0 X_0$$

$$X_0 = \frac{V_{AB}}{I_0} = \frac{182.455}{6} = 30.4096\ \Omega$$

$$X_0 = 2X_0 = 60.818\ \Omega$$

Now of slip of 0.05 me can write

$$Z_f = \frac{J\,X_0\left[\left(\frac{r_2}{5}\right) + J\,x_2\right]}{\frac{x_2}{5} + J(x_2 + x_0)} = \frac{J30.4091\left[\frac{0.3889}{0.05} + J1.9811\right]}{\left[\frac{0.3889}{0.05} + J(1.98.11 + 30.4091)\right]}$$

$$= \frac{30.4091\angle 90° \times 1.99\angle 84.25°}{32.39\angle 89.64°} = 1.868\angle -83.89°$$

$$= 0.1988 - J\,1.8573\ \Omega$$

$$Z_{eq} = Z_1 + Z_f + Z_f$$

$$= 1.8 + J\,3.9622 + 6.4819 + J\,3.416 + 0.1988 - J\,1.8573$$

$$= 8.4807 + J\,5.5209 = 10.1194\ \angle 33.06°\ \Omega$$

$$I_1 = \frac{V_1}{Z_{eq}} = \frac{220\angle 0°}{10.1194\angle 33.06°} = 21.74\angle -33.06° \text{ A}$$

$$\text{P.F.} = \cos(-33.06°) = 0.838 \text{ logging}$$

Now $$V_f = I_1 Z_f = 21.74\angle -33.06° \times \angle 27.79°$$

$$= 159.288\angle -5.27° \text{ V}$$

$$I_{2f} = \frac{V_f}{\left(\frac{r_2}{S} - J\,x_2\right)} = \frac{159.288\angle -5.27°}{\left(\frac{0.3889}{0.05} J\,1.9811\right)} = \frac{159.288\angle -527°}{8.0263\angle 14.28°}$$

$$= 19.8457\angle -19.55° \text{ A}$$

$$V_b = I_1 Z_b = 21.74\angle -33.06 \times 1.868\ \angle -83.89°$$

$$= 40.61\angle -116.95° \text{ V}$$

$$I_{2b} = \frac{V_b}{\left(\frac{r_2}{2-5} + J\,x_2\right)} = \frac{40.61\angle -116.95°}{1.99\angle 84.25°}$$

$$= 20.407\angle -201.2° \text{ A}$$

$$P_f = (I_{2f})^2 \times \frac{r_2}{S} = \frac{(19.8457)^2 \times 0.3889}{0.05} = 3063.379 \text{ W}$$

$$P_b = (I_{2b})^2 \times \frac{r_2}{2-S} = \frac{(20.407)^2 \times 0.3889}{1.95} = 83.05 \text{ W}$$

$$P_m = (1 - S)\,(P_f - P_b) = (1 - 0.05) \times (3063.379 - 83.05)$$

$$= 2831.3125 \text{ W}$$

$$W_0 = \text{rotational lower} = 350 \text{ W}$$

$$P_{out} = P_m - \text{rotational lower} = 2831.3125 - 350$$

$$= 2481.3125 \text{ W}$$

$$P_{in} = V_1 I_1 \cos\phi = 220 \times 21.74 \times 0.838$$

$$= 4007.98 \text{ W}$$

$$\%\eta = \frac{P_{out}}{P_{in}} \times 100 = \frac{2481.3125}{4007.98} \times 100$$

$$= 61.909\%$$

Example 2.2

Find the input current power factor and efficiency of a $\frac{1}{2}$ *hp, 110 V, 50 Hz, single phase induction motor base on double revolving field theory with the following data of a slip of 5%.*

Stator impedance = 2 + *J* 3 Ω

Equivalent rotor impedance = 2 + *j* 3 Ω

Magnetising impedance = 50 Ω

Friction and voltage less = 25 *W*

Solution: The given values are, $R_1 = 2\ \Omega$, $X_1 = 3\ \Omega$

$$R_2 = 2\ \Omega \quad \text{and} \quad X_2 = 3\ \Omega$$

$$r_2 = \frac{R_2}{2} = 1\ \Omega \quad \text{and} \quad X_2 = \frac{X_2}{2} = 1.5\ \Omega$$

$$X_0 = 50\ \Omega \text{ and hence } x_0 = \frac{X_0}{2} = 2.5\ \Omega$$

$$Z_f = \frac{J x_0 \left[\left(\frac{r_2}{S}\right) + J x_2\right]}{\frac{r_2}{S} + J(x_0 + x_2)} = \frac{(J\,25)\left[\left(\frac{1}{0.05}\right) + J1.5\right]}{\frac{1}{0.05} + J(25+1.5)}$$

$$= \frac{25\angle 90° \times 20.056\ \angle 4.289°}{33.2\angle 52.95°} = 15.102\ \angle 41.339°\ \Omega$$

$$= 11.338 + J\,9.975\ \Omega$$

$$Z_b = \frac{J x_0 \left[\left(\frac{r_2}{2-5}\right) + J x_2\right]}{\frac{r_2}{2-5} + J(x_0 + x_2)} = \frac{(J25)\left[\left(\frac{1}{1.95}\right) + J1.5\right]}{\frac{1}{1.95} + J(25+1.5)}$$

$$\frac{25\angle 90° \times 1.585\angle 71.12°}{26.504\angle 88.89°} = 1.495\angle 72.23°\ \Omega$$

$$= 6.456 + J\ 1.423\ \Omega$$

$$Z_{eq} = Z_1 + Z_f + Z_b = 2 + J3 + 11.338 + J9.975 + 0.456 + J1.423$$

$$= 13.794 + J14.398\ \Omega = 19.939\angle 46.22°\ \Omega$$

$$I_1 = \frac{V_1}{Z_{eq}} = \frac{110\angle 0°}{19.939\angle 46.22°} = 5.516\angle{-46.22°}\ \text{A}$$

Input current = 5.516 A

$$\text{P.f} = \cos(-46.22°) = 0.6918 \text{ lagging}$$

$$V_f = I_1 Z_f = 5.516\angle{-46.22°} \times 15.102\ \angle 41.339°$$

$$= 83.302\angle{-4.881°}\ \text{V}$$

$$I_{2f} = \frac{V_f}{\left(\frac{r_2}{5} + J\,x_2\right)} = \frac{83.302\angle{-4.881°}}{20.056\angle 4.289°} = 4.153\angle{-9.17°}\text{ A}$$

$$V_b = I_1 Z_b = 5.516\angle{-46.22°} \times 1.495\angle 72.23°$$
$$= 8.2464\angle 26.01°\text{ V}$$

$$I_{2b} = \frac{V_b}{\left(\frac{r_2}{2-5} + Jx_2\right)} = \frac{8.2464\angle 26.01°}{1.585\angle 71.12} = 5.2027\angle{-45.11°}\text{A}$$

$$P_f = (I_{2f})^2 \times \frac{r_2}{S} = \frac{(4.153)^2 \times 1}{1.95} = 13.881\text{ W}$$

$$P_m = (1 - S)\,(P_f - P_b) = (1 - 6.05)\,[344.948 - 13.881]$$
$$= 314.513\text{ W}$$

$$P_{out} = P_m - \text{frictional loss} = 314.513 - 25 = 419.756\text{ W}$$

$$P_{in} = V_1 I_1 \cos\phi = 110 \times 5.516 \times 0.6918 = 419.756\text{ W}$$

$$\%\eta = \frac{P_{out}}{P_{in}} \times 100 = \frac{289.513}{419.756} \times 100 = 68.97\%$$

Example 2.3

A 250 watt, 230 V, 50 Hz. Single phase capacitor start induction motor has the following constants for the main and auxiliary windings main winding. $Z_m = (4.5 + J\,3.7)\ \Omega$. Auxiliary winding $Z_a = (9.5 + J\,3.5)\ \Omega$. Determine the value of the capacitor that will place the main and auxiliary winding currents in quadrature of starting.

Solution: Let X_C be the reactance of the capacitor connected in the auxiliary winding

$$Z_a = (9.5 + J\,3.5)\ \Omega$$

than $\quad Z_a = 9.5 + J3.5 - J\,X_C = 9.5 - J\,X$

Now $\quad Z_m = (4.5 + J\,3.7)\ \Omega = 5.8258\ \angle 39.42°\ \Omega$

Thus I_m lags behind V by 39.42° since time phase angle *** I_m and I_a has to be 90°. I_a must lead V by an angle given by

$$\phi_a = 90° - 39.42° = 50.58°$$

For auxiliary winding

$$\tan\phi_a = \frac{X}{R}$$

$$\therefore \quad \tan 50.58 = \frac{X}{9.5}$$

$$\therefore \quad X = 9.5 \times \tan 50.58$$

$$\therefore \quad X = 11.5572\ \Omega$$

But we have $X_C - 3.5 = X$

$$X_C = 3.5 + X = 3.5 + 11.5572$$

$$X_C = 15.0572\ \Omega$$

$$P_f = (I_{2f})^2 \cdot \frac{r_2}{S} = (3)^2 \times (62.6) = 563.4 \text{ W}$$

$$P_b = (I_{2b})^2 \frac{r^2}{2-S} = (4.5860)^2 \cdot (1.605) = 33.75 \text{ W}$$

Forward torque $T_f = P_f = 563.4$ synchronous watts.

Backward torque $T_b = P_b = 33.75$ synchronous watts.

$$\text{Net torque } T = T_f - T_b$$
$$= 563.4 - 33.75$$
$$T = 529.65 \text{ synchronous watts}$$

Example 2.4

The main and auxiliary winding impedances of a 50 Hz. Capacitor start single-phase induction motor are:

Main winding $Z_{1m} = 3 + J2.7$

Auxiliary winding $Z_{1a} = 7 + J3$

Determine the value of the capacitor to be connected in series with the auxiliary winding to achieve a phase difference of $\alpha = 90°$ *between the current of the two winding at start.*

Solution: Let X_C be the reactance of the capacitor connected in the auxiliary winding.

$\therefore \quad Z_{1a} = 7 + J3 - JX_C = 7 - J\,X$ i.e. $X = X_C - 3$

and $\quad Z_{1m} = 3 + J2.7 = 4.03608 \angle 41.98° \ \Omega$

Thus without X_C, I_m lags V by 41.98° but angle between I_m and I_a has to be 90° thus I_a must lead V by an angle.

$$\phi_a = 90° - 41.98° = 48.02°$$

for auxiliary winding.

$$\tan \phi_a = \frac{X}{7}$$

$$\therefore \tan (98.02°) = \frac{X}{7}$$

$\therefore \quad X = 7.7797 \ \Omega$

But $\quad X = X_C - 3$

$\therefore \quad X_C = 3 + X$

$$= 3 + 7.7797 = 10.7797 \ \Omega$$

Now $\quad X_C = \dfrac{1}{2\pi fc}$

$$C = \frac{1}{2\pi \times 50\,V \times 10.7797}$$
$$= 295.286 \ \mu\text{F}$$

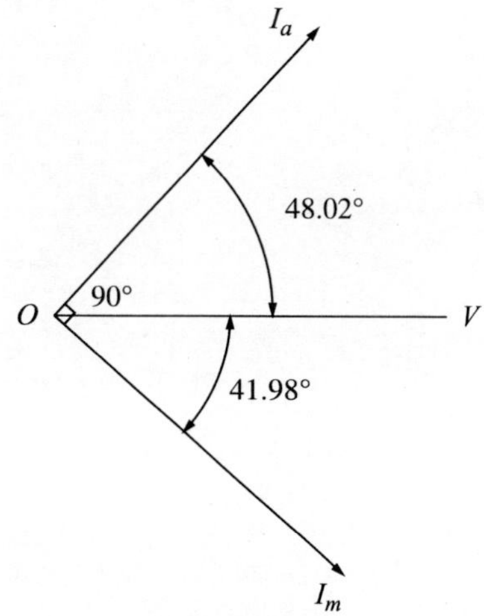

Fig. 2.30

Example 2.5

The following data shown result on a 230 V, 50 Hz. Capacitor start single-phase induction motor of standstill.

Main winding: 100 V, 2 A, 40 W

Auxiliary winding, 80 V, 1.0 A, 50 W

Determine the value of capacitance for attaining the maximum starting torque.

Solution: Z_{1em} = Equivalent impedance with main winding

$$= \frac{V_m}{I_m} = \frac{100}{2} = 50\,\Omega$$

R_{1em} = Equivalent resistance with main winding

$$= \frac{V_m}{I_m^2} = \frac{40}{(2)^2} = 10\,\Omega$$

$$\therefore \quad X_{1em} = \sqrt{(Z_{1em})^2 - (R_{1em})^2} = 48.9897\,\Omega$$

Running rotor resistance and reactance referred to stator R_2' and X_2' equal to stator resistance R_1 and reactance X_1 respectively.

$$R_{1m} = R_2' = \frac{R_{1em}}{2} = 5\,\Omega$$

$$X_{1m} = X_2' = \frac{X_{1em}}{2} = 24.4948\,\Omega$$

$$\phi_m = \tan^{-1}\frac{X_{1m}}{R_{1m}} = 78.46°$$

Z_{1ea} = Equivalent impedance with auxiliary winding

$$= \frac{V_a}{I_a} = \frac{80}{1} = 80\,\Omega$$

R_{1ea} = Equivalent resistance with auxiliary winding

$$= \frac{W_a}{I_a^2} = \frac{50}{(1)^2} = 50\,\Omega$$

$$\therefore \quad X_{1ea} = \sqrt{Z_{1ea}^2 - R_{1ea}^2} = 62.45\,\Omega$$

$\therefore \quad R_a$ = resistance of auxiliary winding = $R_{1ea} - R_2' = 45\,\Omega$

X_a = reactance of auxiliary winding $X_{1ea} - X_2' = 37.9552$

Torque is maximum when angle between I_m and I_a is 90°

$$\phi_a = 90° - 78.46° = 11.54°$$

$$Z_a = R_a + JX_a - JX_c$$

When capacitor is connected

$$\therefore \qquad \phi_a = \tan^{-1}\left[\frac{X_a - X_C}{R_a}\right] = -11.54°$$

ϕ_a is taken negative as I_a leads V by ϕ_a

$$\therefore \qquad \tan(-11.54°) = \frac{37.9552 - X_C}{45}$$

$$\therefore \qquad X_C = 47.1432\ \Omega = \frac{1}{2\pi fc}$$

$$\therefore \qquad C = \frac{1}{2\pi \times 50 \times 47.1432} = 67.52\ \mu\text{F}.$$

❑❑❑

Chapter

3 Stepper Motors and Switched Reluctance Motors

3.1 STEPPER MOTORS

A stepper motor is a digital actuator where input is in the form of programmed energization of the stator winding and where output is in the form of discrete angular rotation.

Rotation occurs because of magnetic interaction between rotor poles and poles of the sequentially energized stator windings. A stepper motor is a brushless DC motor where rotor rotates through a fixed angular step in response to each current pulse received by its controller.

The main feature of a stepper motor is that the rotor rotates in discreate angular intervals or steps, an step being taken each time a command pulse is received. The definite number of pulses are applied to the motor. The rotor rotates through a definite known angle.

Due to this stepper motor is mainly used for open loop position control because no feedback signal from the motor shaft.

- The torque range of stepper motor from (1 M N-m to 40 N-m)
- The output power ranges from (1 W to maximum 2500 W).

In the construction of stepper motor, stator has phase winding. Rotor has no electrical windings, and commulator and brushes, hence this is known as brushless DC motor.

3.2 STEP ANGLE

Step angle is defined as the angle through which the stepper motor shaft rotates for each command pulse, it is denoted as β.

The step angle is very small, the number of steps per revolution is greater. Commonly used step angles are 1.8°, 2.5°, 7.5° and 15°. The minimum value of step angle is 0.72° and maximum value of step angle is 90°. Due to this, higher resolution or accuracy of positioning obtained.

Formula for step angle (β)

(*i*) $$\beta = \frac{(N_s \sim N_r)}{N_s N_r} \times 360^\circ$$

where N_s = Number of stator teeth
N_r = Number of rotor teeth

(*ii*) **Other Formula**

$$\text{Step angle } \beta = \frac{360^\circ}{M.N_r}$$

where M = Number of stator phases

3.2.1 Resolution

The resolution is defined as the number of steps needed to complete one revolution of the rotor shaft.

$$\text{Resolution} = \frac{\text{Number of steps}}{\text{revolution}} = \frac{360^\circ}{\beta}$$

3.2.2 Slewing

The stepper motor may be operate at very high stepping rate *i.e.*, 20,000 steps per second. A stepper motor operates at high speed is called *slewing*.

$$\text{Motor speed } n = \frac{\beta \times f}{360^\circ} \text{ rps}$$

where β = Step angle

f = Stepping frequency or pulse rate in pulses per second (PPS)

When the stepping rate in increased to quickly the stepper motor comes to the out of synchronism and stops.

Advantages

(*i*) It can driven open loop without feedback.

(*ii*) No accumulative position error.

(*iii*) Responds directly to digital control signals, so stepper motors are natural choice for digital computer controls.

(*iv*) It is mechanically simple.

(*v*) It requires little or no maintenance.

(*vi*) Can be repeatedly stated without damage.

(*vii*) Relatively rugged and durable.

Disadvantages

(*i*) Low efficiency with ordinary controller.

(*ii*) Fixed step angle.

(*iii*) No flexibility on step resolution.

(*iv*) High overshoot and oscillation in step.

(*v*) Limited power output and sizes available.

3.2.3 Stepper Motor Application

The main application of stepper motor may be divided into following groups:

1. Instrumentation application
  (*i*) Quartz watches
  (*ii*) Synchronized clocks
  (*iii*) Camera shutter operation

2. Stepper motor application in computer peripherals
  (*i*) Dot matrix and line printers
  (*ii*) Floppy disk drives
  (*iii*) Digital *X-Y* plotter
  (*iv*) Magnetic tape transport
  (*v*) Paper tape drive

3. Office equipment
  (*i*) Electronic type writer
  (*ii*) Facsimile machine

4. Robotics

5. Electromedical application
  (*i*) X-ray machines
  (*ii*) Radiation therapy units
  (*iii*) CAT-Scanners, ultrasound scanners

6. Miscellaneous application
  (*i*) Nuclear reactors
  (*ii*) Aero space.

3.3 CLASSIFICATIONS OF STEPPER MOTORS

These are classified into three types:
  (*i*) Variable Reluctance (VR) stepper motor
  (*ii*) Permanent Magnet (PM) stepper motor
  (*iii*) Hybrid stepper motor.

3.3.1 Variable Reluctance Stepper Motor

The VR stepper motor consists of stator and rotor. The stator windings are wound on the stator poles. The rotor carries no windings. Rotor poles are of a ferromagnetic material. The rotor is a salient poles type. This motor may be single stack type or multistack type. This stepper motor is called *variable reluctance motor* because the reluctance of the magnetic circuit formed by the rotor and stator teeth varies with angular position of the rotor.

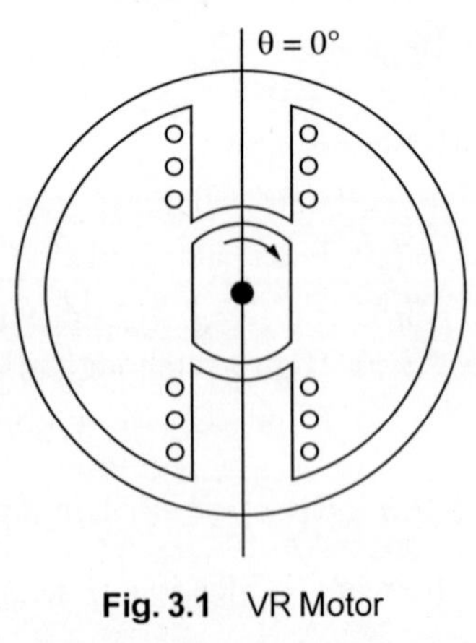

Fig. 3.1 VR Motor

The stator is made of soft iron stamping. It has six equally spaced projecting poles or teeth. The stator winding are wound on stator poles. The rotor is also made of soft iron stamping. It is a 4 pole salient type of the same width as the stator pole or teeth.

The stator winding consists of three phases A, B and C. For the supply current to the stator winding proper sequence shows a circuit in Fig. 3.2.

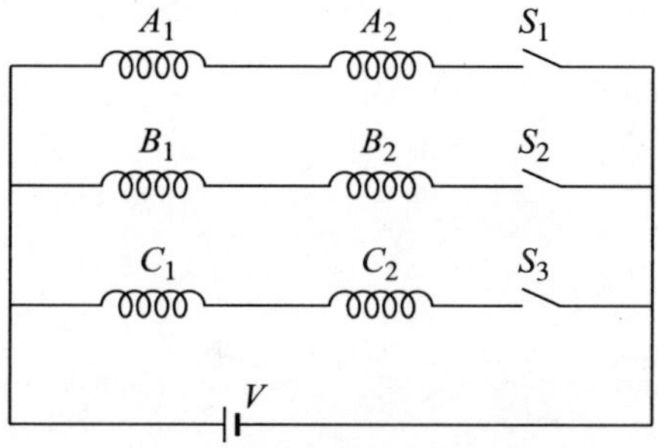

Fig. 3.2

The six stator windings are connected in 2-coil groups to form three separate circuit. It is called *phases*. Phase A consists of two terminals A_1 and A_2. Similarly, Phase B has B_1 and B_2 and Phase C is also connected in C_1 and C_2 with own independent switch. The flow of current can be controlled by mechanical switch or solid-state control. The opposite pairs of stator coil are connected in series such that A_2, B_2, C_2.

When there is no current in stator windings. The rotor is completely free to rotate. By energizing one or more stator windings. The rotor rotates forward or backward to a position that forms a path of minimum reluctance with the magnetized stator poles or teeth.

For three phase, four pole rotor motor. The *step angle* of this motor is

$$\beta = \frac{360^\circ}{MN_r} = \frac{360^\circ}{3 \times 4} = 30^\circ$$

Operation

The VR stepper motor has following modes of operation.

1. 1-phase-ON or full-step operation
2. 2-phase-ON mode
3. Alternate — 1-phase-ON and 2-phase-ON mode (half step operation)
4. Micro-stepping operation

1. Full-step operation or 1-phase-ON mode

1-phase-ON mode means at that time only one winding in energized and remaining winding are de-energized by energizing on stator winding. The rotor rotates by a full step of 90° in clockwise direction.

First we closed switch S_1 and phase A winding is energized. A magnetic field with its axis along the stator poles or teeth of phase A is created. Now the rotor moves to the position on which the rotor teeth align themselves with stator teeth of phase A *i.e.*, teeth 1 and 3 lining up with stator teeth 1 and 4. This is the position in which the reluctance of the magnetic circuit is the minimum. Here $\theta = 0^\circ$.

Next, phase A winding is de-energized by opening switch S_1 and phase B winding is energized by closing switch S_1. Now the rotor teeth 2 and 4 align with stator teeth 3 and 6 in Fig. 3.3 (b).

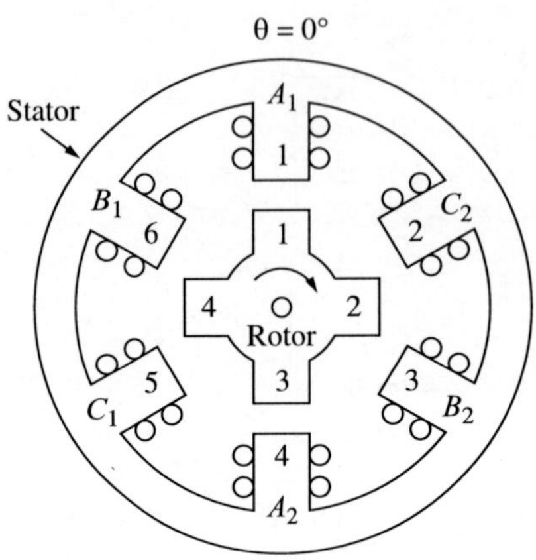

A only
(a)

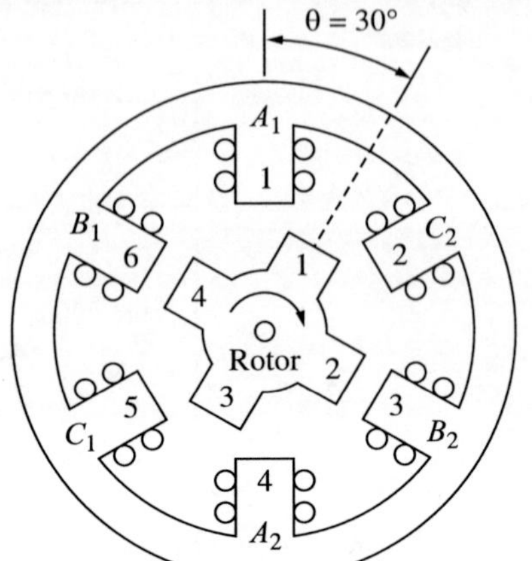

B only
(b)

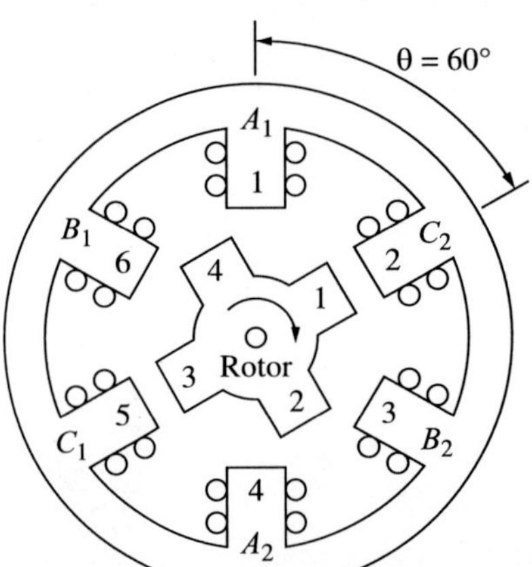

C only
(c)

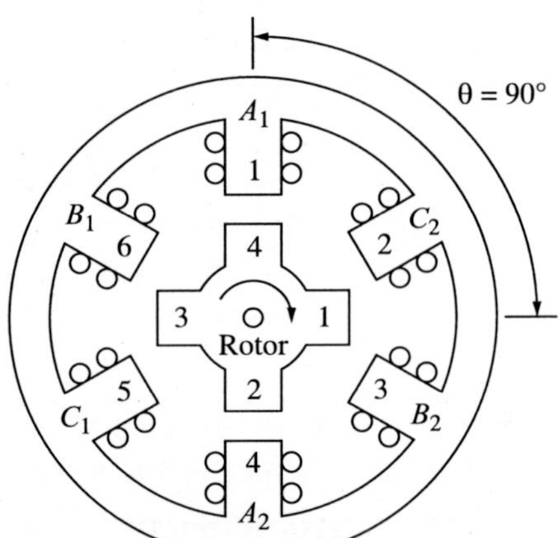

A only
(d)

TRUTH TABLE

Phase A	*Phase B*	*Phase C*	θ
+	0	0	0°
0	+	0	30°
0	0	+	60°
+	0	0	90°

(f)

Fig. 3.3

Here rotor rotates through full step of 30° in the clockwise direction θ = 30°.

Next, phase B winding in de-energized by opening switch S_2 and phase C winding is energized by closing switch S_3. Now the rotor teeth 1 and 3 to align with stator teeth 2 and 5. In Fig. 3.3 (c) here, the rotor rotates another 30° in the clockwise direction, here θ = 60°.

Again phase A winding is energized and phase C winding is de-energized. Now the rotor teeth 2 and 4 to align with stator teeth 1 and 4. In Fig. 3.3 (d) Here, the rotor rotates another 30° in clockwise direction. Here θ = 90°.

The stator winding in energized in the sequence of ABCA. The rotor each time rotates through an angle of 30° in clockwise direction. Fig. 3.3(e) shows the truth table of 1-phase ON mode.

If the sequence is made *CBAC*, the rotor will rotates opposite direction. The direction of the stator magnetizing current is not significant because a stator pole of either magnetic polarity will always attract the rotor pole by inducing opposite polarity.

2. 2-phase-ON Mode

In this mode of operation, two stator winding, are energized simultaneously. The two phase windings (*AB*, *BC*, *CA*, *AB* etc.) are energized the rotor experiences torques from both phases and comes to rest of a point mid-way between the two adjacent full-step positions.

First, phase A and B winding are energized and rotor rotates in the clockwise direction. Here θ = 15°. It is shown in Fig. 3.4(*a*) below:

Next, phase B and C windings are energized and rotor rotates another 30° in clockwise direction. Here θ = 45°. It is shown in Fig. 3.4(*b*).

Next, phase C and A winding are energized and rotor rotates another 30° in clockwise direction. Here θ = 75°. It is shown in Fig. 3.4(c).

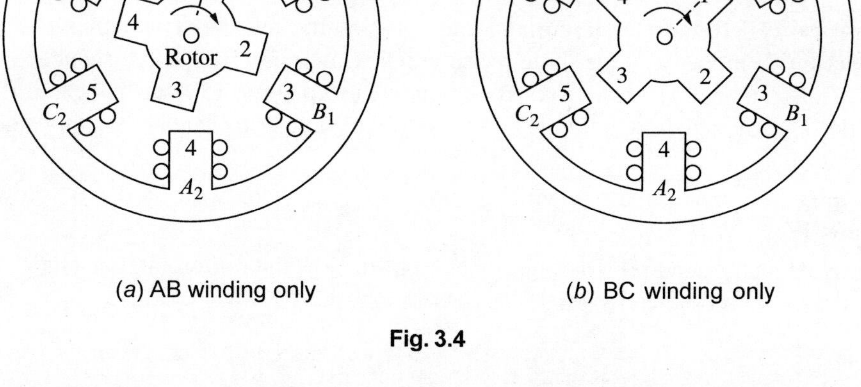

(*a*) AB winding only

(*b*) BC winding only

Fig. 3.4

Next, phase A and B winding are energized and rotor rotates another 30°. In clockwise direction. Hence $\theta = 105°$ it is shown in Fig. 3.4(d).

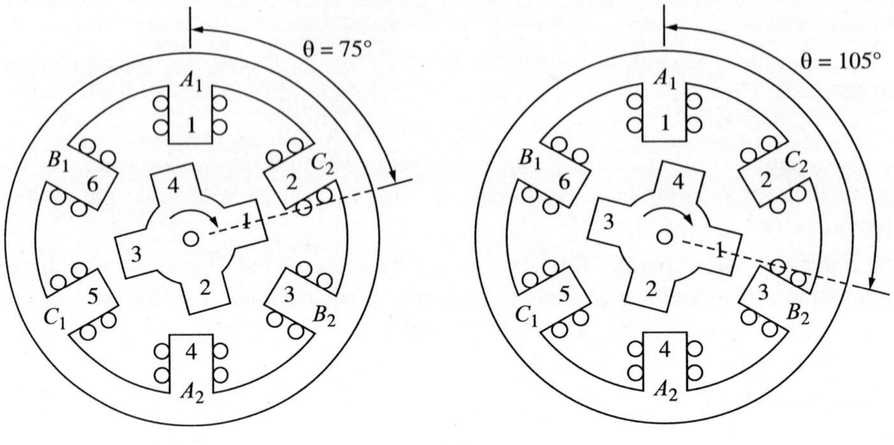

(c) CA only

(d) AB only

Fig. 3.4

Show truth table for 2-phase-ON mode. This mode provides greater holding torque and much better damped single-slack response than the 1-phase on mode operation.

Phase A	*Phase B*	*Phase C*	θ
+	+	0	15°
0	+	+	45°
+	0	+	75°
+	+	0	105°

(f)

Fig. 3.4

3. Half-step operation

It is also called half-stepping operation, it is nothing but alternate 1-phase-ON and 2-phase-ON modes *i.e.*, the three phase winding are energized in the sequence of A, AB, B, BC, C, etc. Here the rotor rotate on each step angle 15° *i.e.*, half the full-step angle. This mode of excitation is known as wave excitation. That is why this mode of operation is called half-step operation.

First, we excite phase A winding only. Now the rotor position $\theta = 0°$ shown in below Fig. 3.5(a).

Next, phase A and B windings are energized. Now the rotor rotates an angle 15° in clockwise direction. Here $\theta = 15°$ shown in Fig. 3.5(b).

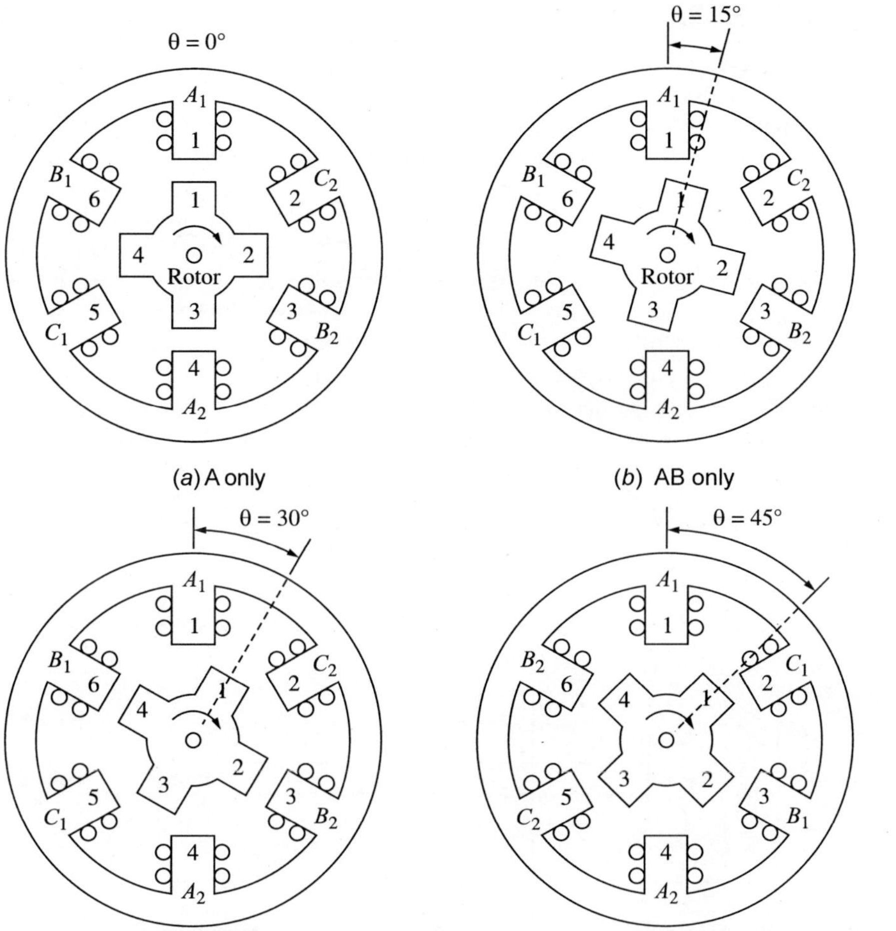

(*a*) A only (*b*) AB only (*c*) B only (*d*) BC winding only

Fig. 3.5

Next, phase *B* winding is only energized and phase *A* winding is de-energized. Now, the rotor rotates another 15° in the clockwise direction. Here $\theta = 30°$ shown in Fig. 3.5(*c*).

Next, phase *B* and *C* winding are energized. Now, the rotor rotates another 15° in clockwise direction. Hence = 45° in clockwise direction shown in Fig. 3.5(*d*).

The truth table for half-step operation. In this mode of operation the step angle is halved *i.e.*, 15°. This is reason resolution is doubled and smooth shaft rotation is achieved.

Truth Table

Phase A	*Phase B*	*Phase C*	θ
+	0	0	0°
+	+	0	15°
0	+	0	30°
0	+	+	45°
0	0	+	60°
+	0	+	75°
+	0	0	90°

3.3.1.1 Microstepping

For resolution of VR stepper motor in 200 steps revolution *i.e.*, $B = 1.8°$. But microstepping have a revolution of 20,000. Steps/revolution. *i.e.*, $\beta = 0.018°$. Microstapping means, the step angle of the VR stepper motor is very small. It is also called mini-stepping. It can be achieved by two phases. Simultaneously as in 2-phase-ON mode but with the two current deliberately made unequal. The current flow through phase *A* winding is held constant and the current flow through phase *B* winding is increased in very small increments until maximum current is reached. Then the current in phase *A* winding is reduced to zero using the same very small increments. In this way, we can get a very small angle it is called microstep. AVR stepper motor with microstepping is mainly used where very fine resolution is required. The application are printing and photo type setting. AVR stepper motor with microstepping provides very smooth low-speed operation and high resolution.

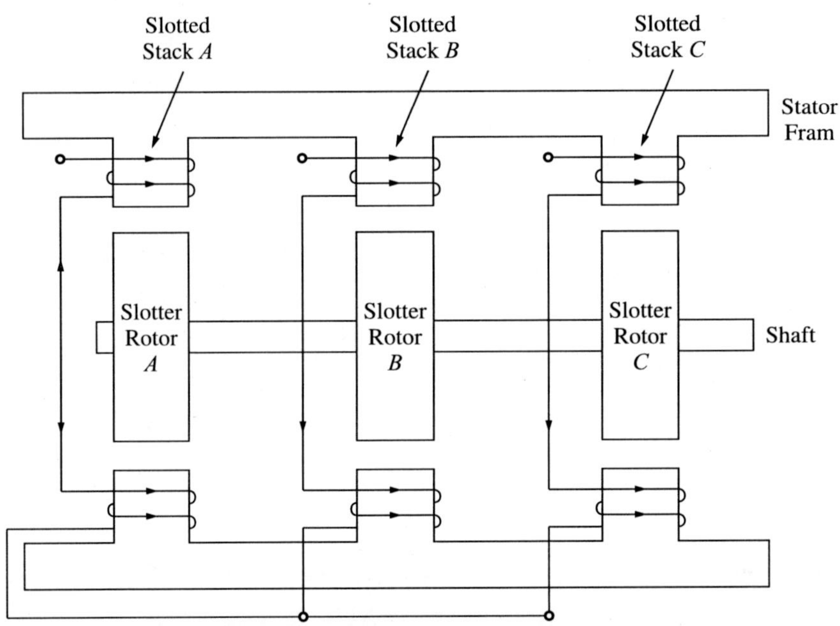

Fig. 3.6 Multistack VR Stepping Motor

3.3.1.2 Multistack VR Stepping Motor

Another technique of microstepping is to use a multistack VR stepping motor. It is more common to use a three stack VRSM.

3.3.1.3 Construction

It has three separate magnetically isolated section or stacks shown in Fig. 3.6. The stator has common frame but stator laminations for each stack are displaced axially from each other see Fig. 3.6.

- The stator laminations are also offset in angular displacement by one-third (for a 3 stack motor) of a stator pole pitch for a 12-pole stator. The pole pitch is 30° and therefore, the three stator stacks must be offset from each other in angular displacement of 10° as shown in Fig. 3.9.

- In Fig. 3.7(*a*) is shown stack *A* with stator teeth or poles 1, 2, 3, 4 ... stack *B* is shown in Fig. 3.7(*b*) with stator poles 1, 2, 3, 4, ... displaced from stack *A* by 10° CCW. Further in Fig. 3.7(*c*) the stack *C* as shown in with its stator poles 1, 2, 3, 4 ... offset from stack *B* by an angular displacement of 10° CCW.
- Note that rotor teeth on the three stacks do not have any angular displacement from each others as the stator poles have, the rotor teeth are aligned.

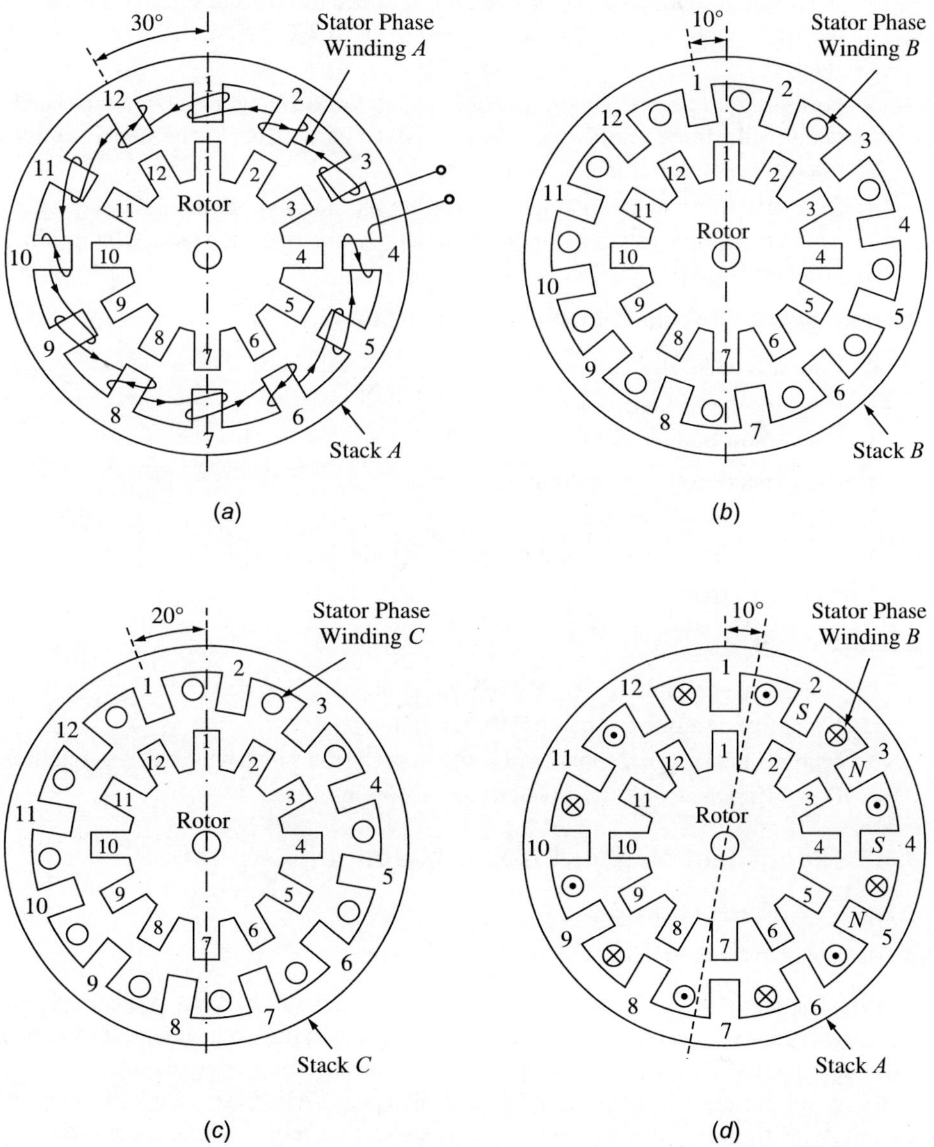

Fig. 3.7

3.3.1.4 Operation

1. When phase winding on stator stack A is excited, alternate N and S poles are produced on stator Fig. 3.7(*a*). The rotor teeth tend to align with the stator poles so that reluctance seen by the flux of stator stack A is minimum, this is shown in Fig. 3.7(*a*).
2. When current in stator winding of stack A is reduced to zero and stack B is energized, then rotor moves through a step of 10° CCW. So as to minimize the reluctance seen by the stack flux B. This is shown in Fig. 3.7(*d*).
3. If current in stator winding or stack B is reduced to zero and stack C is energized, the rotor would rotate further by a step of 10° CCW with respect to stator stack B.
4. Further, with de-energization or stack C winding and reenergization of stack A winding the rotor would again none by 10° CCW to execute one step to complete the rotation through an stator pole pitch of 30°.
5. CCW rotation can be obtained by exciting the stator winding on the sequence of *ABCABCA*, CW rotation can be obtained by exciting the stator winding in the sequence of *ACBA CBA*.

Advantages of Variable Reluctance Stepper Motor

1. Low motor inertia
2. High torque to inertia ratio
3. Capable of high stepping rate
4. High speed slowing capability
5. High weight
6. 3, 4 and 5 phase, single and multi-stack models available
7. Ability to freewheel.

Disadvantages

1. Normally available in 36° to 30° step angles.
2. No definite torque available with windings de-energized.
3. Exhibits mid-range resonance of some stepping rates under some drive conditions.
4. Low efficiencies of low voltages and stepping rates.

3.3.2 Permanent Magnet Stepper Motor

3.2.2.1 Construction

It consists of two main parts, one is stator and other one is rotor.

Stator: The stator is made up of a stack of steel laminations. It has projecting poles. The stator has 4 poles. The stator windings or phase windings are wound. In the stator poles here the two stator poles are energized by one winding. Now the motor has two windings or phases. It is marked as A and B. The phase A consists of two terminal *i.e.*, A_1 and A_2. Similarly phase B winding consists of two terminal *i.e.*, B_1 and B_2.

Rotor: The rotor is a permanent magnet. It is made up of magnetically 'hard' ferrite. The motor is cylindrical one, here the rotor consists of two poles.

$$\text{Step angle } \beta = \frac{360°}{MN_r}$$

$$M = 2, N_r = 2$$

$$\beta = \frac{360°}{2X_2} = 90°$$

3.3.2.2 Operation

1-pole ON mode: One of the stator winding is energized, the rotor poles move into alignment with the energized stator poles. The stator windings can be excited with either polarity current. A^+ refers to positive current I_A^+ flow through the phase winding A. A refers to negative current I_A^- flow through the phase winding A.

First we consider phase winding A is energized with positive current I_A^+. It is shown in Fig. 3.8. here angle rotation is zero then the phase B winding is energized it is shown in Fig. 3.8(a). Now rotor rotates 90° in the clockwise direction *i.e.*, $\theta = 90°$. Now, the phase winding A is energized with negative current I_A^- *i.e.*, current flows from A_2 to A_1.

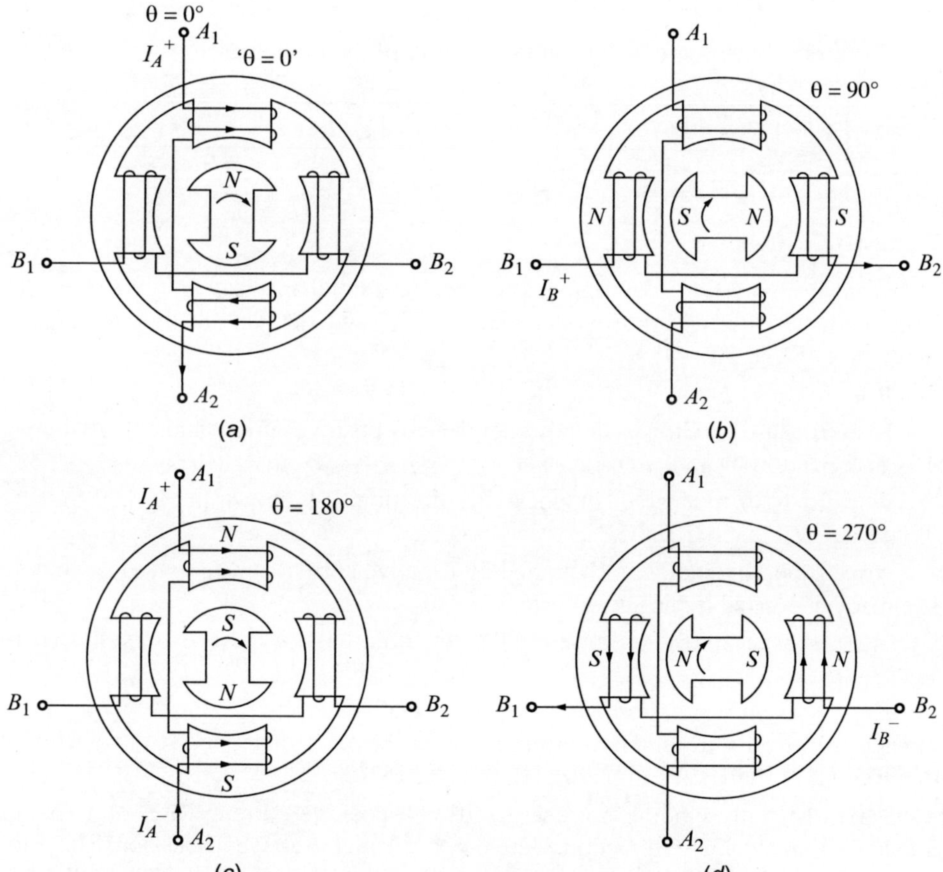

Fig. 3.8

Then the rotor rotates through another 90° in clockwise direction. It is shown in Fig. 3.8(*c*), where $\theta = 180°$.

Now, the phase winding *B* is energized with negative current I_B^- *i.e.*, current flow from B_2 to B_1.

Then the rotor rotates through another 90° in clockwise direction. Here $\theta = 270°$, it is shown in Fig. 3.8(*d*).

Next, the energization of phase winding *A* with I_A^+ makes the rotor rotates through one complete revolution of 360°. The above operation is single-phase ON mode. The truth table Fig. 3.8 (*e*) shows single phase ON mode.

Phase A	*Phase B*	*Rotation* θ
+	0	0°
0	+	90°
–	0	180°
0	–	270°
+	0	0°

The motor direction depends on the polarity of the phase currents.

Clockwise direction	$I_A^+, I_B^+, I_A^-, I_B^-, I_A^+, I_A^+$
Counter clockwise direction	$I_A^+, I_B^-, I_A^-, I_B^-, I_B^+, I_A^+$

2-phase-ON mode

In this mode of operation, two windings are energized simultaneously first. We energized both stator winding with positive current *i.e.*, I_A^+ and I_B^+. Due to this the angular rotation is 45° *i.e.*, $\theta = 45°$.

It is shown in Fig. 3.9(*a*).

Next, *B* phase winding is already energized by positive current and phase winding *A* is energized by negative current *i.e.*, I_A^-.

Now the rotor rotates 90° in clockwise direction. It is shown in Fig. 3.9(*b*), here $\theta = 135°$.

Next, *B* phase winding is energized by negative current *i.e.*, I_B^- phase *A* winding is already energized by negative current.

Now, the rotor rotates another 90° in clockwise direction it is shown in Fig. 3.9(*c*). here $\theta = 225°$.

Next, phase *A* winding is energized, with positive current I_A^-. Phase *B* winding is already energized with negative current I_B^-. Now the rotor rotates and 90° in clockwise direction. It is shown in Fig. 3.9(*d*). Here $\theta = 315°$.

Next, the both winding are energized with positive current. The rotor rotates another 90° in clockwise direction. Here $\theta = 45°$ from the above discussion. The rotor pole positions are midway between the two adjacent full-step position. The truth table (*e*) shows 2-phase on mode operation.

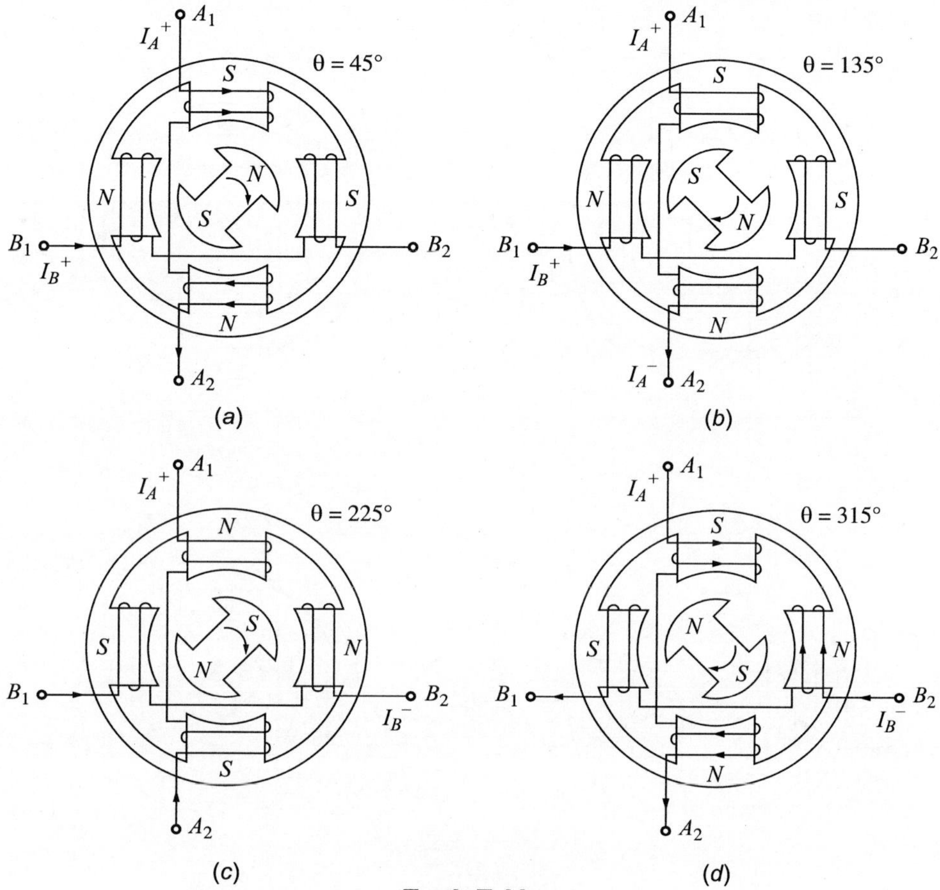

Truth Table

Phase A Winding	*Phase B Winding*	θ
+	+	45°
–	+	135°
–	–	225°
+	–	315°
+	+	45°

(*e*)

Fig. 3.9

Alternate 1-phase ON and 2-phase ON Mode

In the mode of operation, first only winding is energized and rotor rotates at 45°, after that two windings are energized and again rotor rotates 45°, the detail operations are given below.

First, phase *A* winding only is energised with positive current *i.e.*, I_A^+. Now the rotor rotation angle is 0. *i.e.*, $\theta = 0°$ it is shown in Fig. 3.10(*a*). Next, the *B* winding is energized by positive current *i.e.*, I_B^+ phase. A winding is already energized with I_A^+. Now the rotor rotates 45° in clockwise direction. It is shown in Fig. 3.10(*b*), here $\theta = 45°$.

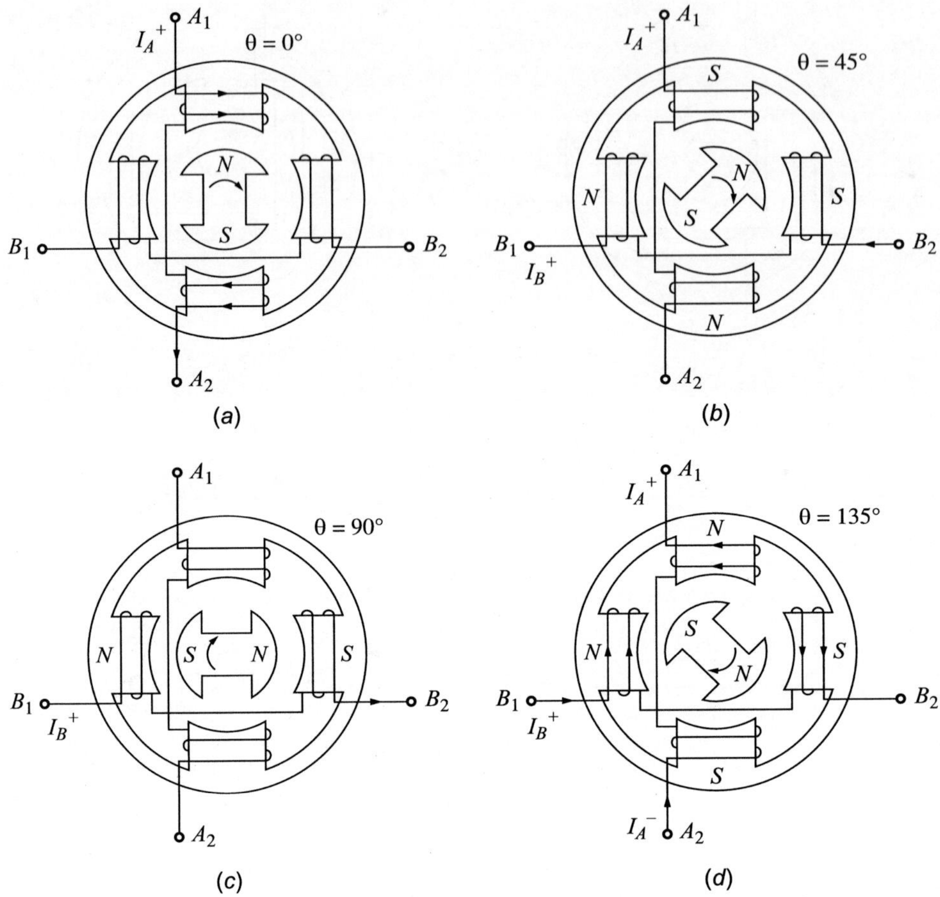

Fig. 3.10

Next, phase A winding is de-energized and phase B winding is already energized with positive current I_B^+. Now, the rotor rotates 45° in clockwise direction. It is shown in Fig. 3.10(c), here $\theta = 90°$.

Next, phase A winding is energized with negative current I_A^-. Phase B is already energized with positive current I_B^+. Now, the rotor rotates 45° in clockwise direction. It is shown in Fig. 3.10(d), here $\theta = 135°$.

Next, phase B winding is de-energized and phase A winding is already energized with negative current I_A^-. Now the rotor rotates 45° in clockwise direction.

It is shown in Fig. 3.10(c). Here $\theta = 180°$. Next both winding are energized by negative current I_A^- and I_B^-. Then the rotor rotates 45° in clockwise direction, here $\theta = 225°$, it is shown in Fig. 3.10(f).

Next, phase A winding is de-energized and phase B winding is already energized with negative current I_B^-. Now the rotor rotates 45° in clockwise direction. It is shown in Fig. 3.10(g), here $\theta = 270$.

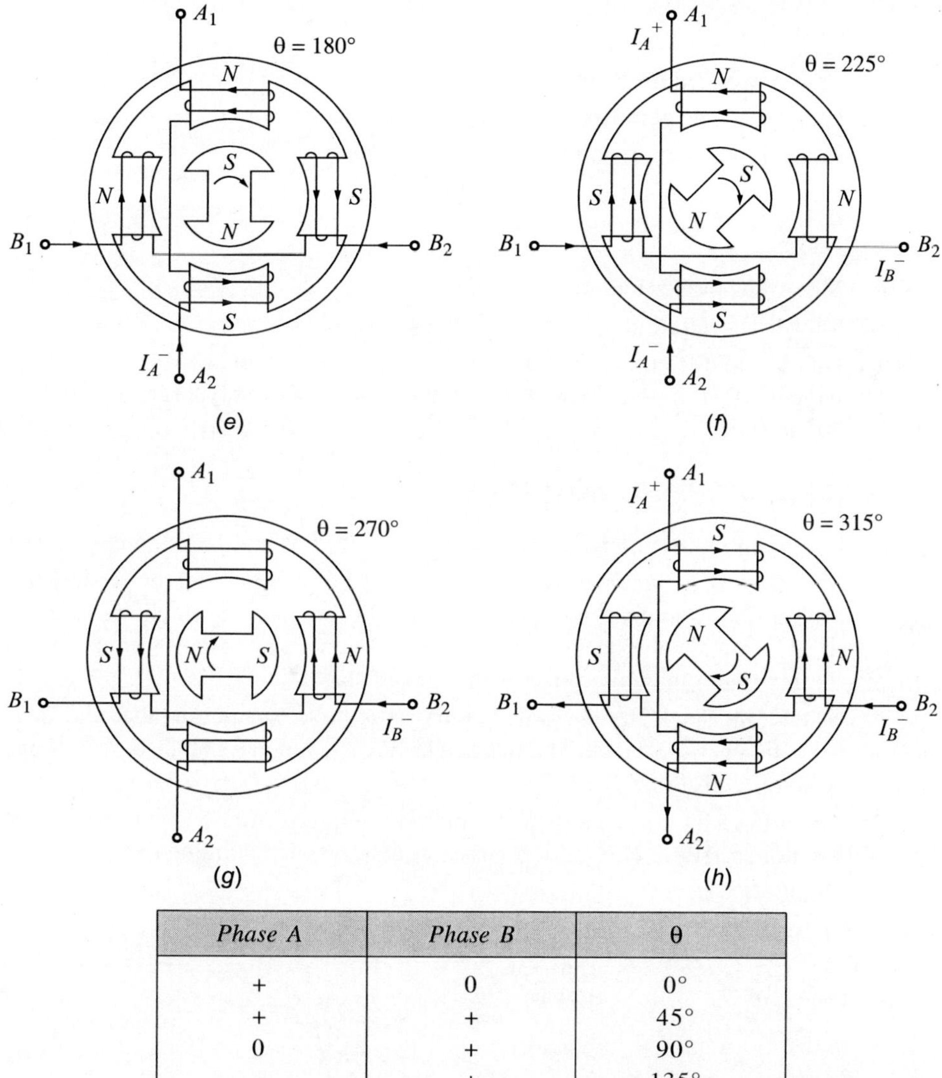

(e) (f) (g) (h)

Phase A	*Phase B*	θ
+	0	0°
+	+	45°
0	+	90°
–	+	135°
–	0	180°
–	–	225°
0	–	270°
+	–	315°
+	0	0° or 360°

(*i*)

Fig. 3.10

Next, phase *A* winding is energized with positive current I_A^+, and phase *B* winding is already energized by negative current I_B^-. Now the rotor rotates 45° in clockwise direction, here θ = 315° it is shown in Fig. 3.10(*h*). Next, phase *B* winding is de-energized and phase *A* winding is already energized with positive current I_A^+. Then the rotor rotates 45° in clockwise direction, it is shown in Fig. 3.10(*a*), here θ = 0° or 360°. From the above operation we can form the truth table Fig. 3.10(*i*).

Advantages

(*i*) Low power requirement.
(*ii*) High torque as compared to variable reluctance motor.
(*iii*) Rotor do not require external exciting current.
(*iv*) It produces more torque per ampere stator current.

Disadvantage

(*i*) Motor has higher inertia.
(*ii*) Slower acceleration.
(*iii*) Due to the difficulties of manufacture a small permanent magnet rotor with large number of poles. The step size in motors is relatively large ranging from 30° to 90°.

3.4 HYBRID STEPPING MOTORS

A hybrid stepping motor combines the features of both PM and VR stepping motors.

The constructional view of the rotor of a hybrid stepping motor is shown in Fig. 3.11 it consist of an axial permanent magnet of the ends of which are attached. Two identical ferromagnetic stacks as shown. These two stacks consists of equal number of teeth, in this figure there are three teeth on each stack.

At one end, the stack attains north (N) magnetic polarity and the other end, stack gets south (S) magnetic polarity. The two stacks have an angular displacement of one-half of the rotor teeth pitch.

The stator has salient-pole structure which is continues from one end to the other end of the stator structure. The stator poles carry concentrated winding like other types of stepping motors.

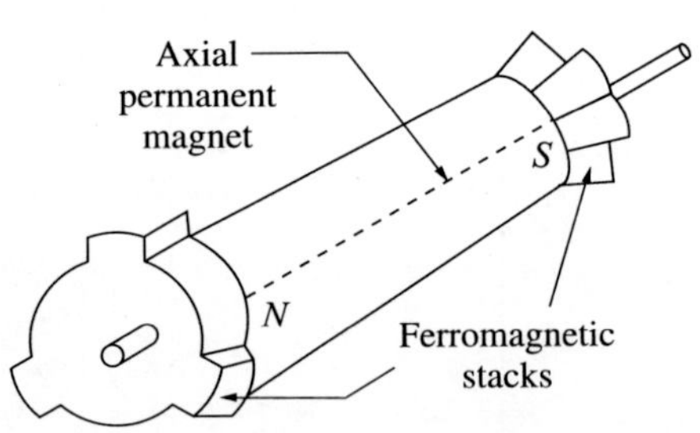

Fig. 3.11

Operation

For understanding the operation of hybrid stepper motor (HSM) refer to Fig. 3.12(*b*). In this figure north poles of the front end of Fig. 3.12(*a*) are shown with full lines whereas the south pole of the far end are shown with dotted lines.

When phase winding it is energized with current I_a, N pole at A_1 and S pole at A_2 are created on the stator pole at A_1 attracts S pole of far end and pole of A_2 attracts N pole of front end, this is shown in Fig. 3.12(*a*). This equilibrium positions of rotor structure results in maximizing. The flux linkages with the phase winding A. Here rotation $\theta = 0°$.

For turning the rotor clockwise through a step, de-energize phase winding A and excite phase winding B so that N pole at B_1 and S pole at B_2 are created on stator.

Pole at B_1 attracts S pole of rear end and pole at B_2 attracts N pole of front end. So a step angular rotation of 30° clockwise (CW) is achieved Fig. 3.12(*b*). In this equilibrium position maximum flux linkage are now linked with phase winding B.

If excitation is removed from phase winding B and reversed excitation is applied to phase winding A. Pole on A, attracts front N pole and pole of A_2 attracts rear S pole. This gives rise to a further step angle movement of 30° CW, Fig. 3.12(*c*). In this manner 12 steps will complete are revolution.

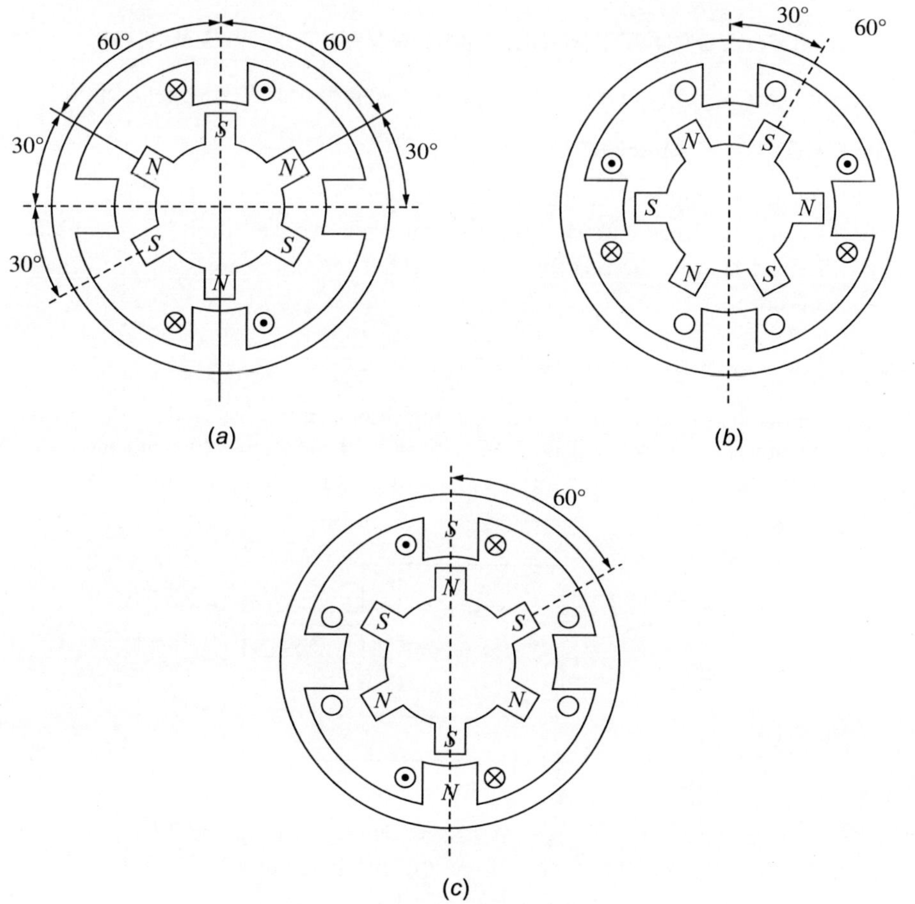

Fig. 3.12

Sequence of exciting the phase winding for clockwise rotation in $A_1\ B_1\ A_2\ B_2\ A_1$ and therefore of CCW rotation, the sequence will be $A_1\ B_2\ A_2\ B_1\ A_1$.

As in $PMSMS_1$ the rotor poles in HSM also tend to align themselves with the stator teeth when stator exciting current are absent.

Advantages of Hybrid Stepper Motor

1. Provide detente torque with winding de-energised.
2. Less tendency to resonate.
3. High stepping rate capability.
4. High efficiency at lower steads and lower stepping rates.
5. High holding torque capability.
6. Better damping due to presence of rotor magnet.

Disadvantage

1. Higher inertia and weight due to presence of rotor magnet.
2. Performance affected by change in magnet strength.

3.5 THEORY TORQUE PRODUCTION IN VR STEPPING MOTOR

VR motor always works in a way to reduce reluctances of the magnetic circuit.

Basic Field Theory Approaches

1. Interms of magnetic energy and co-energy.
2. Considering ideal case of rotor and stator cores have infinite permeability.
3. Proceed step-by-step to the case in which the cores are subject to magnetic saturation.

3.5.1 The Case of Infinitely Permeable Cores

To analyse the situation of an iron piece being drawn into a magnetic field created by an electromagnet as shown in Fig. 3.13. A current I is flowing through the coil of N turns to field magnetic flux, and a force f is acting on the iron piece in the X-direction.

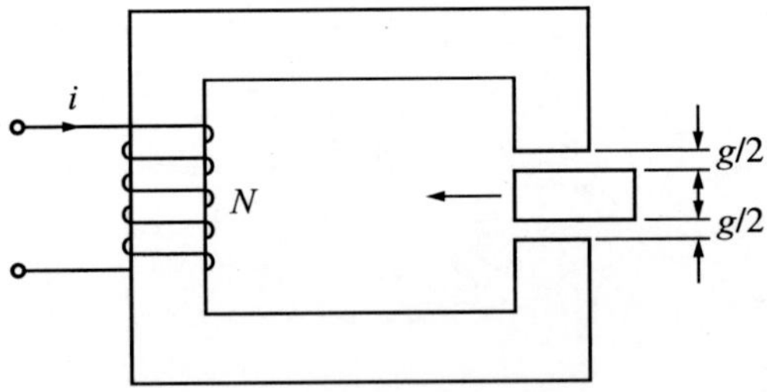

Fig. 3.13

Here, the iron piece may be regarded as a forth of the rotor of a stepping motor the electromagnet corresponds to a pair of teeth of the stator in a VR motor.

Let, B_g = magnetic flux density in the airgaps.

According to Ampere's law along the dotted closed loop is expressed as

$$\oint H.dl = nI$$

$$\oint Hdl = H_g\left(\frac{g}{2}\right) + H_g\left(\frac{g}{2}\right) + Hi(l)$$

$$= H_g g + H_i(l)$$

where H_g = Magnetic field in the gap.

H_i = Magnetic field intensity in the cores.

l = Total magnetic path in the cores.

Where the permeability of cores is extremely large, and $H_i = 0$.

So,
$$\oint H\,dl = H_g \cdot g$$
$$H_g \cdot g = nI$$
$$H_g = \frac{nI}{g}$$

The airgap flux density
$$B_g = \frac{\mu_0\, nI}{g}$$

where μ_0 = Permeability in the gap length.

Let

ω = Transverse length of the iron piece.

x = Distancc by which the rotor tooth and the iron piece overlap.

Overlapped area = $x\omega$

$$\text{Magnetic flux } \phi = x\omega \frac{\mu_0\, nI}{g}$$

$$\text{Flux leakages } \psi = n\phi = x\omega \frac{\mu_0 n^2 I}{g}$$

Now let us assume that there is an incremental displacement Δx of the tooth during a time interval Δt. Then the increment in the flux linkage $\Delta\psi$ is
$$\Delta\psi = \frac{\omega\mu_0 n^2 I}{g} . \Delta x$$

The emf induced in the coils by the change in flux linkage is
$$e = -\frac{\Delta\psi}{\Delta t} = \frac{\omega\mu_0 n^2 I}{g} \cdot \frac{\Delta x}{\Delta t}$$

The minus sign in this equation implies that the direction of the emf is opposing the current. Since the current I is supplied by the power source for the time internal Δt over coming the back emf. The workdone ΔP_i by the source is
$$\Delta P_i = I/e/\Delta t = \frac{\omega M_0 h^2 I^2}{g} \Delta x$$

For analysis purpose, the coil resistance is zero, using equation (1) ΔP_i is expressed interms of B_g as follows.
$$\Delta P_i = \frac{B_g^2}{\mu_0} g\omega\Delta x \qquad ...(1)$$

The workdone by the source is converted partly to mechanical work and the rest is spent increasing the magnetic field energy in the gaps.

The increase in the gap field energy is given by

$$\Delta\omega_n = \frac{1}{2}\frac{B_g^2}{\mu_0} x_i \text{ (the increase in the gap space)}$$

$$= \frac{1}{2}\frac{B_g^2}{\mu_0} g\omega\Delta x \qquad ...(2)$$

From Eqn. (1) and (2), we can find that a half of ΔP_i is converted in to the magnetic field energy in the gaps and other half of ΔP_i is converted into the mechanical work. Since the mechanical work is the force f multiplied by the displacement Δx

$$\Delta W_m = f\,\Delta x$$

$$f\,\Delta x = \frac{1}{2}\frac{B_g^2}{\mu_0}(g\omega\Delta x) \qquad ...(3)$$

Eliminating Δx from both sides

$$f = \frac{1}{2}\frac{B_g^2}{\mu_0} g\omega \qquad ...(4)$$

Here I know $B_g = \dfrac{\mu_0 nI}{g}$

Putting the value of B_g in Eq. (4)

$$f = \frac{1}{2}\frac{\omega\mu_0 h^2 I^2}{g} \qquad ...(5)$$

On the other hand, the magnetic energy w_m in the gap is

$$W_m = \frac{1}{2}\frac{B_g^2}{\mu_0} g\omega$$

From Eqn. (4) and (5)

$$f = \left(\frac{d\omega_m}{dx}\right)_{1 = \text{constant}} \qquad ...(6)$$

The Eqn. (6) must be described in the rigorous form

$$f = \left(\frac{d\,\omega_m}{dx}\right)_{1 = \text{constant}} \qquad ...(7)$$

The equation is valid for the general core in which the coil resistance is not zero.

$$f = -\left(\frac{d\omega_m}{dx}\right)_{1 = \text{constant}}$$

3.5.2 The Case Constant Permeabilities

In the infinitely permeable cores, the magnetic field appears only in the gaps and the mathematical treatment is simple. When cores are of finite permeability, on the other hand magnetic energy appears not only in the gaps. But also in the cores and spaces other than the gaps.

It is not easy to analysis such situations by means of electromagnetic field theory. Instead we will derive an expression for force in terms of circuitry parameters under the assumption that the permeabilities are not the function of magnetic field.

From the Fig. 3.13,

$$L = \text{Coil inductance}$$

$$\psi = \text{Flux linkage}$$

$$\psi = LI$$

The magnetic energy

$$W_m = \frac{1}{2} LI^2$$

emf induced in the coil

$$e = -\frac{\Delta\psi}{\Delta t}$$

$$= \frac{-\Delta(LI)}{\Delta t} \quad \text{...(8)}$$

If the power supply is a current source and provides a current I during the displacement Eq. (8) is simplified as

$$e = -I\frac{\Delta L}{\Delta t} \quad \text{...(9)}$$

Work ΔP_i

$$\Delta P_i = I/e/\Delta f$$

$$= I\left|-I\frac{\Delta L}{\Delta f}\right|\Delta f$$

$$= I^2 \Delta L \quad \text{...(10)}$$

On the other hand, the increase in the magnetic energy ΔW_m in

$$\Delta W_m = \frac{1}{2} I^2 \Delta L \quad \text{...(11)}$$

From Eqn. (10) and (11) it is noted that only half of the workdone by the source is converted to magnetic energy and other half is converted to mechanical work ΔP_0.

$$\Delta P_0 = f\,\Delta x$$

$$f\,\Delta x = \frac{1}{2} I^2 \Delta L \quad \text{...(12)}$$

$$f = \frac{1}{2} I^2 \frac{\Delta L}{\Delta x} \quad \text{...(13)}$$

In the above procedure it was assumed that the coil resistance was zero and the power supply was a current source.

3.6 EFFECTS OF MUTUAL INDUCTION

The number of drive schemes for stepping motors are:

(*i*) One-phase on drive.
(*ii*) Two-phase on drive.
(*iii*) Three-phase on drive.
(*iv*) Half step drive.

In a drive scheme other than the one-phase- on drive it is desirable that the mutual inductance is minimum. Since the mutual inductance has a general tendency to degrade the accuracy in positioning. When mutual inductance is not negligible the torque in terms of linear theory is derived by the following procedure. Consider this Fig. 3.14 the power sources are current sources.

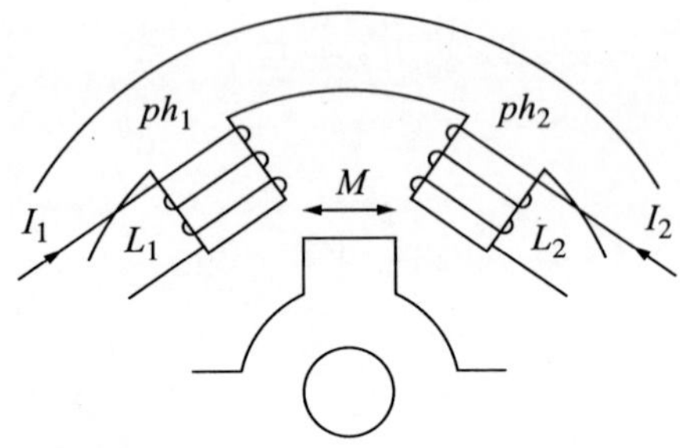

Fig. 3.14

The induced voltages of each phase are:

$$e_1 = -I_1 \frac{\Delta L_1}{\Delta t} - I_2 \frac{\Delta M}{\Delta t} \qquad ...(14)$$

$$e_2 = -I_2 \frac{\Delta L_2}{\Delta t} - I_1 \frac{\Delta M}{\Delta t} \qquad ...(15)$$

where e_1 = Induced voltage of phase 1
e_2 = Induced voltage of phase 2
L_1 = Induced of phase 1
L_2 = Induced of phase 2
M = Mutual inductance between the two phases.

The workdone by the two power supplies during the time increment Δt is

$$\Delta P_i = -(e_1 I_1 + e_2 I_2)\ \Delta t$$

$$= I_1^2 \Delta L_2 + I_2^2 \Delta L_2 + 2 I_1 I_2 \Delta M \qquad ...(16)$$

On the other hand the increment of the magnetic energy on the system is

$$\Delta W_m = \frac{1}{2}(I_1^2 \Delta L_1 + I_1^2 \Delta L_2) + I_1 I_L \Delta M \qquad ...(17)$$

From Eqn. (16) and (17) it is seen that a half of the work done by the sources is converted to an increment of the magnetic energy. The other half is converted into the mechanical output

$$\Delta P_0 = T\,\Delta\theta$$

$$T\Delta\theta = \frac{1}{2}(I_1^2 \Delta L_1 + I_2^2 \Delta L_2) + I_1 I_2 \Delta M$$

Hence torque

$$T = \frac{1}{2} I_1^2 \frac{dL_1}{d\theta} + \frac{1}{2} I_2^2 \frac{dL_2}{d\theta} + I_1 I_2 \frac{dM}{d\theta}$$

3.7 CHARACTERISTICS OF STEPPER MOTOR

Stepper motor characteristics are divided into two groups.

(*i*) Static characteristics
(*ii*) Dynamic characteristics

3.7.1 Static Characteristics

(*i*) Torque-angle curve
(*ii*) Torque-current curve

Torque-angle Curve

A curve is drawn between torque and angle it is shown in Fig. 3.15(*a*).

From the curve, torque increases almost sinusoidally. With angle θ from equilibrium position.

Holding torque is the maximum load torque which the energized stepper motor can withstand without stepping from equilibrium position it is denoted as T_H.

If the holding torque is exceeded, the motor suddenly strips from the prevent equilibrium position and goes the next stable equilibrium position. Figure 3.15(*b*) shows PM and hybrid stepper motor torque and detente torque properties.

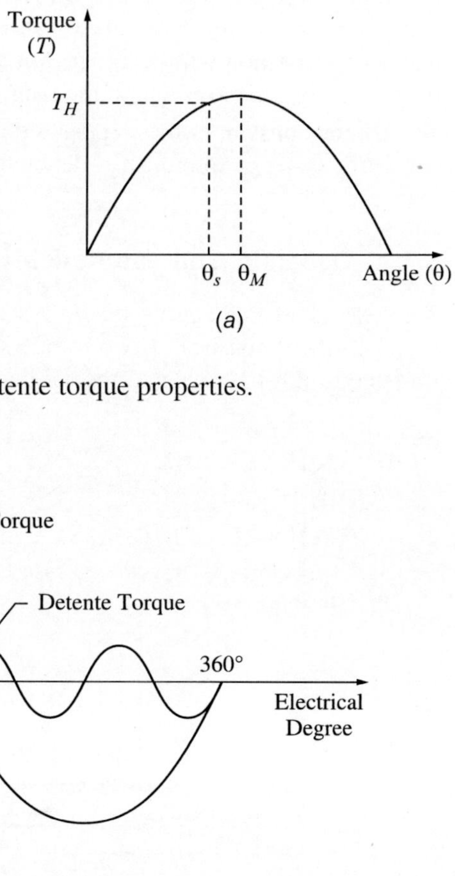

(*b*)

Fig. 3.15

Detente torque is the maximum load torque which is energized stepper motor can withstand stepping it is denoted as T_D.

Detente torque occurs to the residual magnetism present in the stepper motor. It is available only in PM and hybrid stepper motors. It is about 5-10% of holding torque. It is typically a fourth harmonic torque. It is shown in Fig. 3.15(*b*). This torque is also called as cogging torque.

Torque-Current Curve

A curve is drawn between torque and current Fig. 3.15(*c*) shows typical torque-current curve for a stepper motor.

Here the current is zero, some torque will be present in the motor. It is the detente torque again we increasing the current torque linearly increases. The current reaches at rated value. The motor produces holding torque (T_H). Now, the current increases above rated value magnetic saturation will be occur. Initially the curve is linear, but later on, its slope progressively decreases due to the magnetic saturation in the motor.

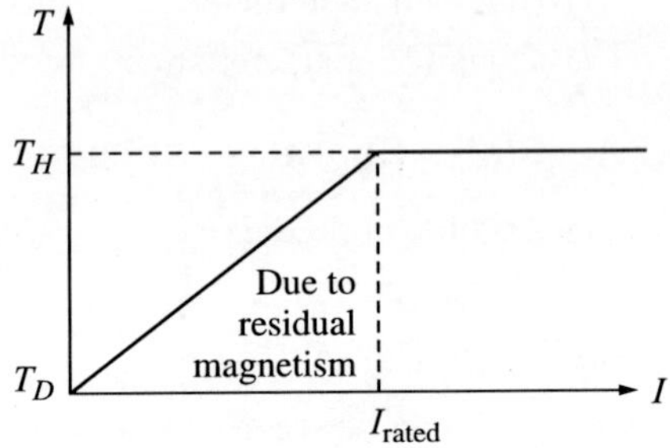

Fig. 3.15(*c*)

Torque constant of the stepper is defined as the initial slope of the torque-current curve of the stepper motor. It is denoted as k_t. It is also called as torque sensitivity its units are N-m/A, kg-Cm/A.

3.7.2 Dynamic Characteristics

This characteristics can be drawn between torque and stepping rate it is shown in Fig. 3.15(*d*). It consists of two torque stepping rate curves as shown in Fig. 3.15(*d*) first one is pull-in curve and second one is pull-out curve.

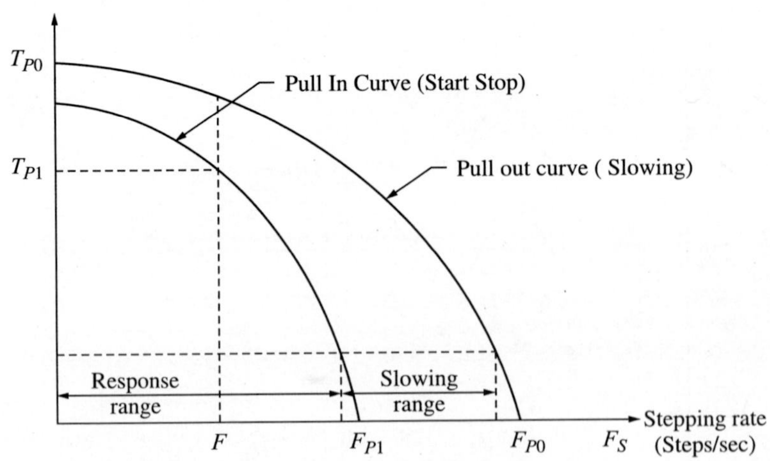

Fig. 3.15(*d*)

Pull-in Curve

Pull-in curve is also called as start-stoper single stepping mode of stepper motor operation. It is shown in Fig. 3.15(*e*).

In this mode of operation, the rotor comes to rest after moving through one step. Consequently, the motor will not move any further as soon as you stop energizing the motor winding in the given programmed sequence. The motor will be start or stop depends upon the respond to each individual pulse.

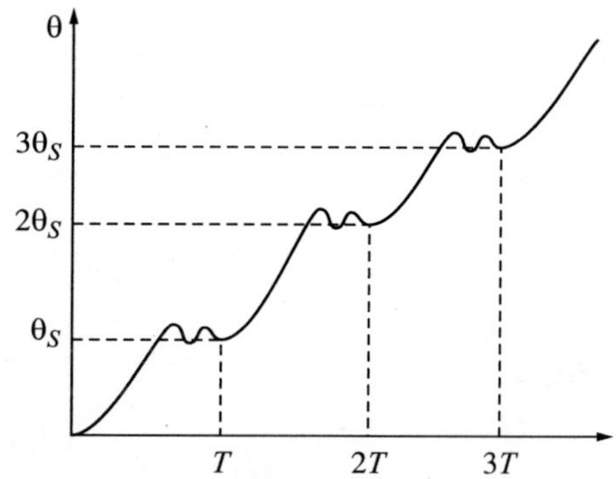

Fig. 3.15(*e*)

Pull-out Curve

It is also called as slewing mode of stepper motor operation. It is shown in Fig. 3.15(*e*).

In this mode of operation, the rotor is still moving in response to the previous pulse when the next pulse comes. Consequently, the motor can be rotate faster rate in slewing mode than in start-stop mode.

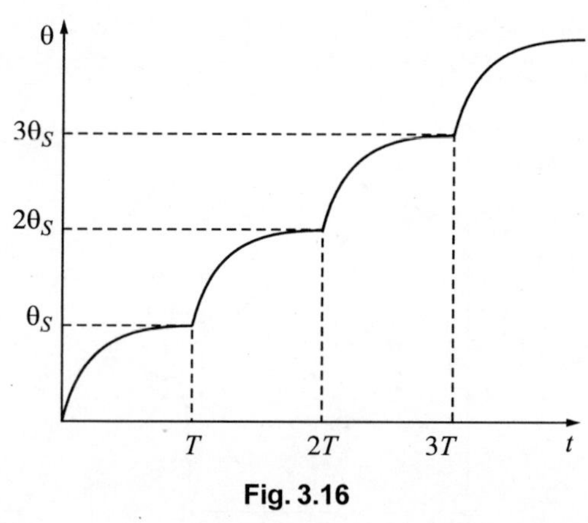

Fig. 3.16

However, the motor cannot start slewing from rest, nor can it stop immediately when you stop applying pulses. The motor will, in fact over run by several steps before it comes to rest.

The following important terms of the motor of pull-in and pull-out courses shown in Fig. 3.15(*d*).

Pull-in Torque (T_A)

It is the maximum torque. The stepper motor can develop in start-stop mode of a given stepping rate E_S (steps/sec) without taking synchronism, it is denoted as T_{P1}.

Pull-in Rate (F_{P1})

It is the maximum stepping rate at which the stepper motor will start or stop, without losing synchronism, against a given load torque T. It is denoted as F_{P1}.

Pull-out-Rate (F_{P0})

It is the maximum stepping rate at which the stepper motor will slew without losing synchronism against a given load torque *T*. It is denoted as F_{P0}.

Response Range (RR)

It is the range of stepping rates at which the stepper motor can start or stop without synchronism, at a given load torque *T*. Response range span stepping rates $F_S < F_{PI}$ the pull in rate.

Stew Range (SR)

It is range of stepping rates at which the stepper motor can run in the stewing mode, with losing synchronism at a given load torque *T*, the slewing range spans stepping rates $F_{P1} < F_S < F_{P0}$.

Synchronism

It is the one-to-one correspondence between the number of pulses applied to the stepper motor controller and the number of steps through which the motor has actually moved. It is importance in the design of control systems employing stepper motors.

Mid-frequency Resonance

In the pull-in-curve of a stepper motor, suddenly dips very low in a particular range of stepping rates, this phenomenon is known as mid-frequency resonance. It is shown in Fig. 3.17.

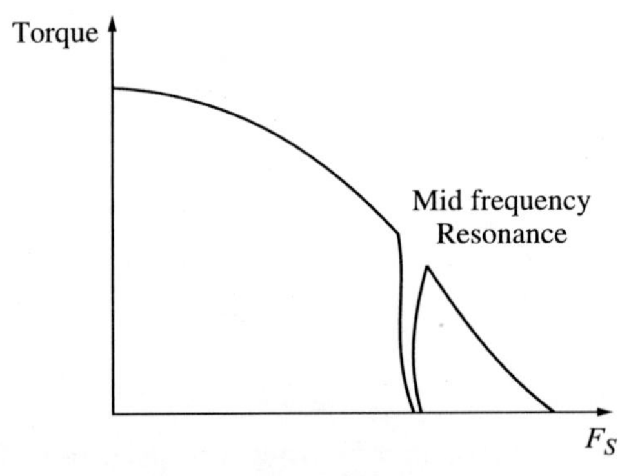

Fig. 3.17

This phenomenon is a manifestation of instability motor operation.

3.8 POWER DRIVE CIRCUITS

The output from the logic sequence generator signals are low level signals which are too weak to energize stepper motor windings. To increase the voltage, current and power levels of the logic sequence output by using switching power amplifier (power semiconductor switching circuits).

The output of the switching power amplifiers are sufficiently high. This output is fed to the stepper motor circuits. The power switching amplifier circuits are called power drive circuits. Some of the power drive circuit are discussed below.

3.8.1 Simple Power Drive Circuit

Figure 3.18(a) shows power transistor driver circuit for one phase of a stepper motor.

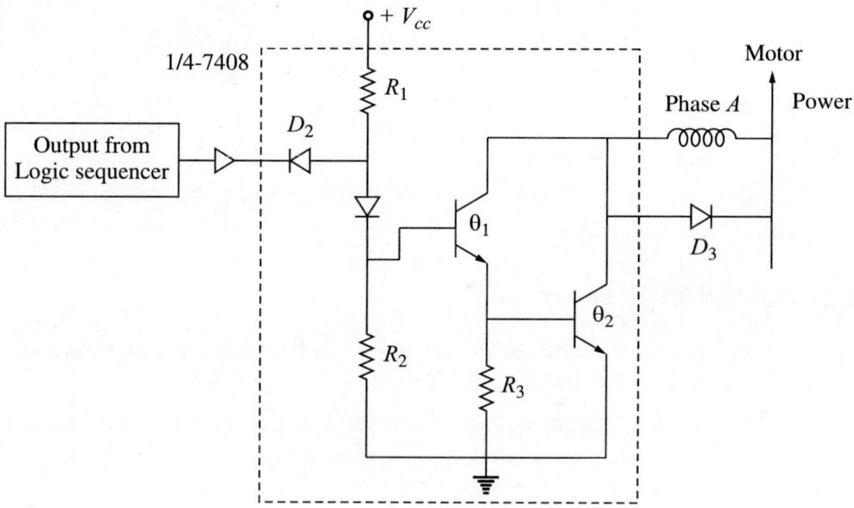

Fig. 3.18(a)

Here, the power circuit is a Darlington pair power amplifier, it is capable of driving 3 to 5 A current through the stepper motor windings.

An inverter gate (1/4 – 7408) is connected between logic sequence and Darlington pair circuit, it is mainly used to isolation between logic sequencer and power driver. This isolation is very important to protect the logic sequencer against any faults. e.g., short circuits, open circuits switching transients, etc.

A free wheeling diode D_3 is connected across motor winding it is used to maintain continuously of current through the motor winding when it is turned off.

L/R Drive

Figure 3.18(b) shows L/R drive circuit.

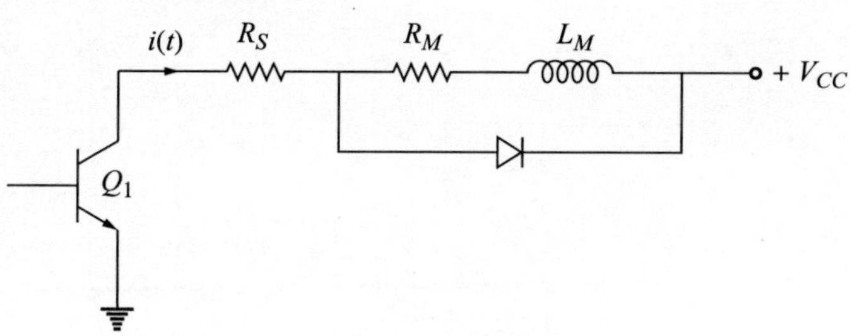

Fig. 3.18(b)

Here R_M and L_M are motor winding resistance and inductance. A diode D is connected across motor winding. It is used for the wheeling purpose.

In this drive circuit a resistance R_S is connected in series with the motor windings it is shown in Fig. 3.18(b), higher input voltage is applied to the circuit.

The motor winding time constant is reduced to $T_e' = \dfrac{L_M}{R_M + R_S}$. Due to the reduction of electrical time constant, the stepper motor current rises much more rapidly and hence the motor develop greater torque at high stepping rates.

Drawbacks

1. Power loss occur in the R_S.
2. L/R drive circuit is used to small instrument stepper motor where rated current is a few hundred MA/phase.

3.8.2 Dual Voltage Drive

To reduces the power dissipation in the driver and increase the performance of a stepping motor, the dual voltage driver is used.

Figure 3.18(*c*) shows dual voltage drive for one phase. When a step command pulse is given to the logic sequences a high-level signal will be put out from one of the output terminals to excite a phase winding.

On this signal both θ_1 and θ_2 are turned on and higher voltage V_H with be applied to the winding. The diode D_1 is now reverse-biased to isolate the lower voltage supply from the higher voltage supply. The current rises up quickly due to the higher voltage V_H.

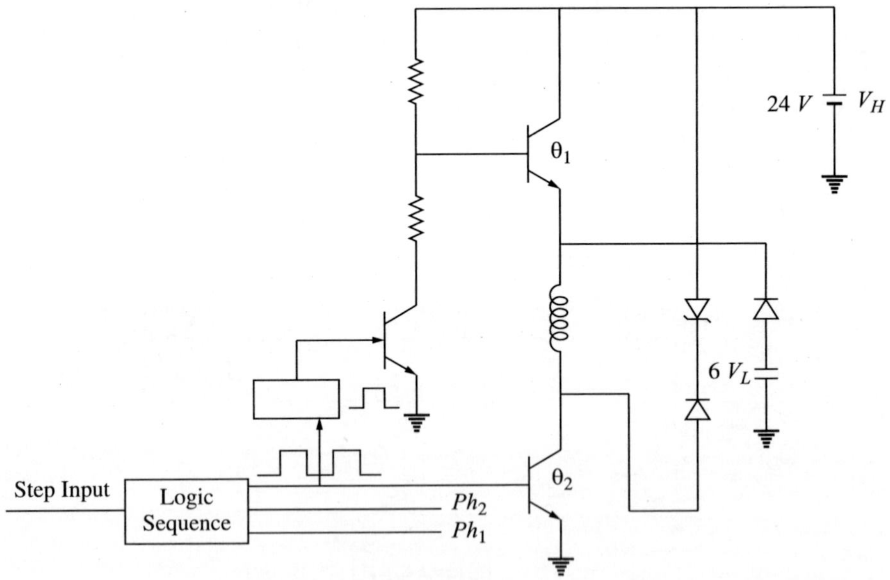

Fig. 3.18(*c*)

The time constant of the monostable multivibrator is selected so that transistor θ_1 is turned off when the winding current exceeds the rated current by a little.

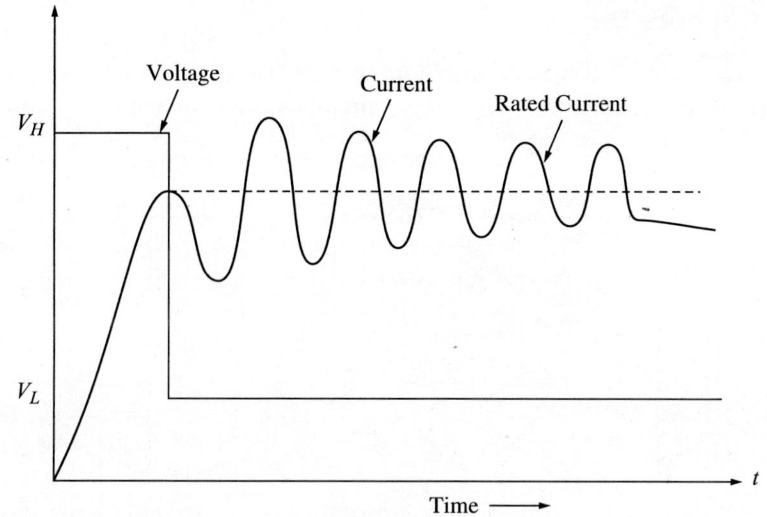

Fig. 3.18(*d*)

After the higher voltage source is cut-off, the diode is forward biased and the winding current is supplied from the lower voltage supply. A typical voltage current waveform are shown in Fig. 3.18(*d*).

Figure 3.18(*e*) shows dual-voltage driver for the two-phase on drive of a four phase motor. Two transistors are used for switching the higher voltage in order that the winding once fed from thc higher voltage is not again excited by it when a new phase is turned on.

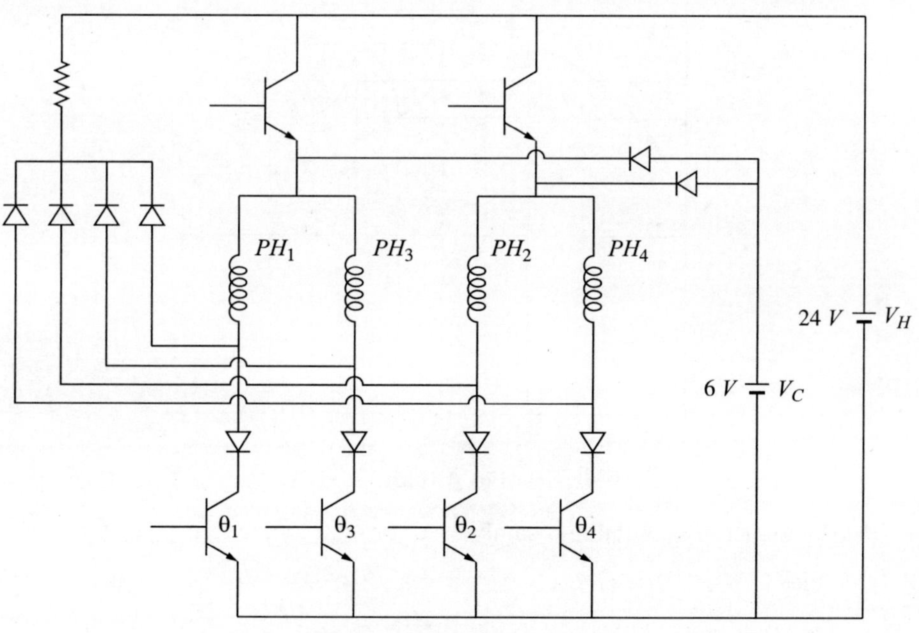

Fig. 3.18(*e*)

3.8.3 Chopper Drive

Figure 3.18(*e*) shows basic chopper drive circuit. Here, the circuit consists of one power transistor *Q* and current sense logic circuit. A high voltage $V_H = (4 \text{ to } 10)\ V_R$ is applied to the motor winding.

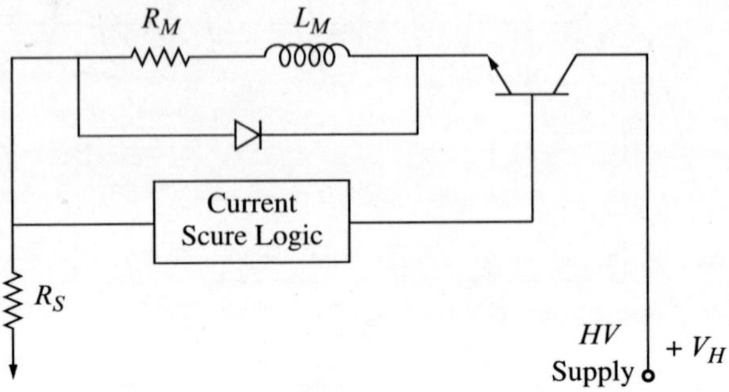

Fig. 3.18(*f*)

When the transistor *Q* is turned *Q* is turned on, the motor current increases. When the motor current maximum value (I_{max}) the transistor *Q* is turned off. Now the current free wheels through diode and decreases exponentially.

Now, the current reaches minimum value (I_{min}) again the transistor *Q* is turned on, this causes current to oscillate maximum and minimum value, it is shown in Fig. 3.18(*f*). Due to the large value of winding inductance, motor current becomes constant *i.e.*, rated current I_R.

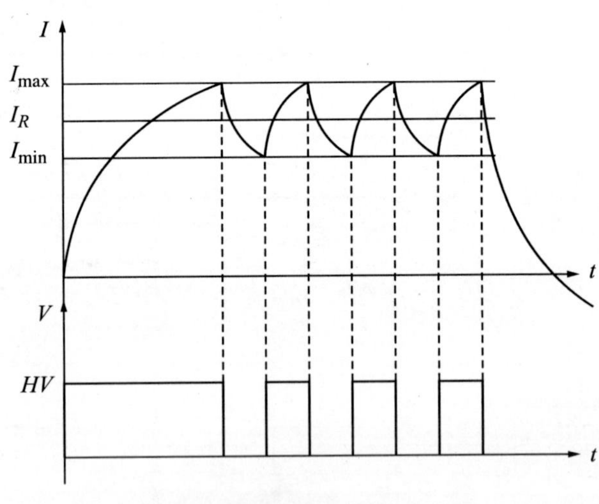

Fig. 3.18(*g*)

This chopper drive circuit is mainly used for high torque stepper motors.

Advantages

1. Energy efficient
2. Simple circuit.

3.8.4 Bipolar Drives for Stepper Motors

So, for we discussed about unipolar drive circuits. In unipolar drive circuits, the current in any particular. Motor winding flows in one direction only this means that the current flow through the motor winding is either rated current (I_R) or zero (0) but it never reverses its direction of flow.

Another classification of drive circuit is bipolar drives by using this drive circuit, the current through the motor winding will be either positive current ($+I_R$) or negative current ($-I_R$) and zero (0). Such drive circuits which cause bidirectional current flow through the motor windings. That is why this drive circuits are called bipolar drives.

Some of the bipolar drives are discussed below.

(*i*) Basic bipolar drives
(*ii*) Bipolar L/R drives
(*iii*) Bipolar chopper drives.

1. Basic Bipolar Drives

Figure 3.19(*a*) shown in below basic bipolar drive circuit. This drive circuit consists of two transistors, two diodes and two power supply.

The stepper motor winding *A* is connected through transistor Q_1 and Q_2 to a bipolar DC source having output voltages $+V$, 0, $-V$.

When the transistor Q_1 is turned on motor current ($+I$) flows through winding *A* *i.e.*, A_1 to A_2 it is shown in below.

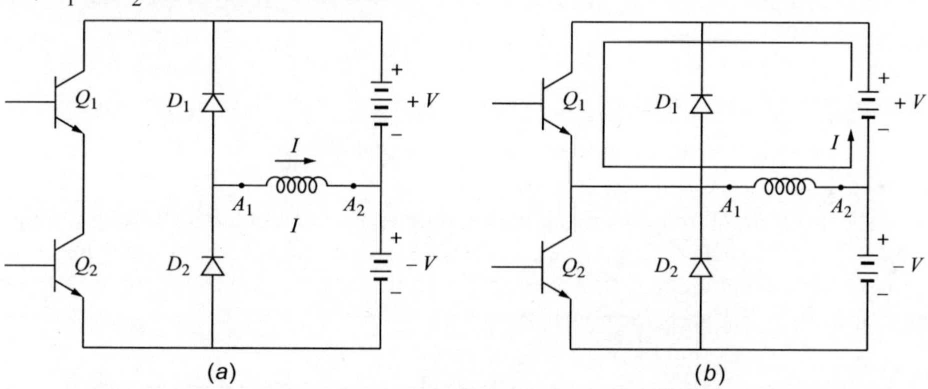

Fig. 3.19

When the transistor Q_1 is turned off and transistor Q_2 is turned on motor current ($-I$) flows through winding *A* i.e., A_2 to A_1, it is shown in Fig. 3.20. In this basic bipolar drive circuit the current is bi-directional. Here D_1 and D_2 are free wheeling diodes.

Figure 3.21 shows practical circuit for bipolar drive of a 4 lead PM hybrid stepper motor. It consists of two stator windings *A* and *B*.

The circuit diagram consists of four transistors two power supply and two phase windings.

The transistor Q_1 and Q_3 control the phase winding *A*, when the transistor Q_1 is turned on, the current flow through the winding *A* is positive *i.e.*, A_1 and A_2 when the transistor Q_3 is turned on and Q_1 is turned off. The current flow through the winding

A is negative *i.e.*, A_2 and A_1 transistor Q_4 and Q_2 control the phase winding B, when the transistor Q_4 is turned on. The current flow through the winding B is positive *i.e.*, B_1 to B_2 when the transistor Q_2 is turned on and transistor Q_4 is turned off.

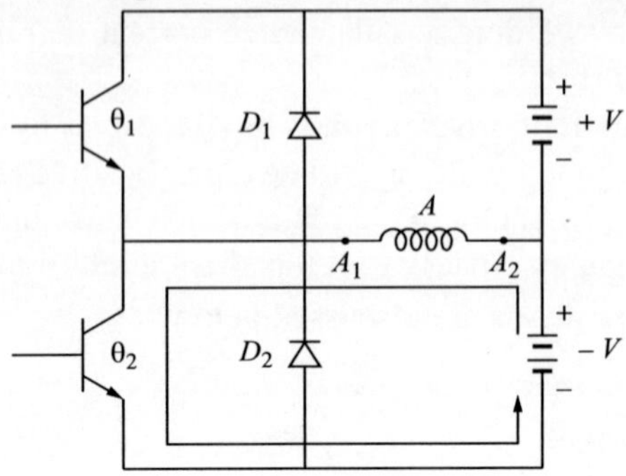

Fig. 3.20

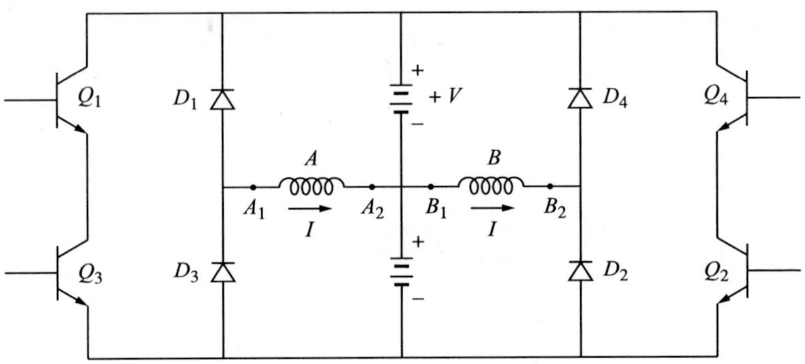

Fig. 3.21

The current flow through the winding B is negative *i.e.*, B_2 to B_1. Bi-directional current flow this takes place.

The drive circuit requires two power supply is bipolar regulated DC power supply with + V, – V outputs. It is possible to obtain bi-directional current flow using a one power supply i.e., unipolar. Power supply with outputs + V and 0. Figure 3.22 shows bipolar drive using unipolar power supply.

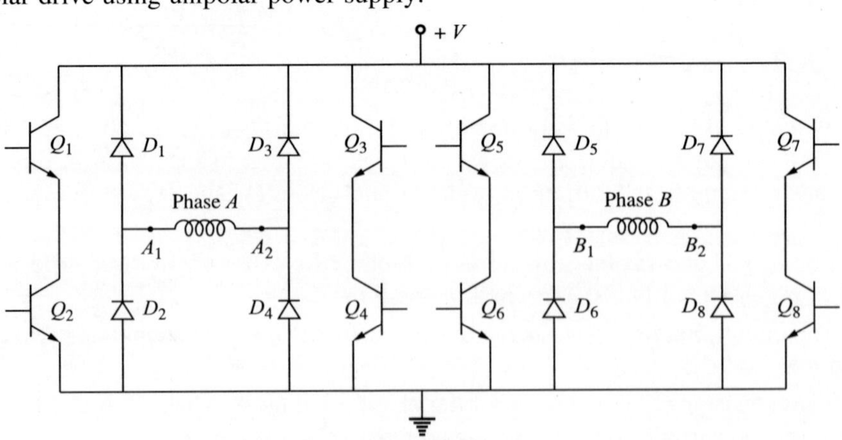

Fig. 3.22

The circuit consists of eight transistor (Q_1 to Q_8) two phase winding *A* and *B* and one power supply.

Here each motor winding requires for transistors form *H* circuit, that is why it is called *H*-type bipolar drive using unipolar power supply. When the transistor Q_1 and Q_4 is turned on the current flow through the motor winding *A* is positive *i.e.*, A_1 to A_2 (+*I*). When the transistors Q_3 and Q_2 is turned on the current flow through the phase winding *A* is negative *i.e.*, A_2 to A_1 (–*I*).

Similarly transistor Q_5 and Q_8 is turned on the current flow through the motor winding *B* is positive *i.e.*, B_1 to B_2 (+*I*). When the transistor Q_7 and Q_6 is turned on the current flow through the motor winding *B* is negative *i.e.*, B_2 to B_1 (–*I*). Here, the bi-directional current takes place by using unipolar power supply.

Bipolar L/R Drives

Figure 3.23 below shows basic bipolar L/R drive. This circuit is similar to Fig. 3.23 except that resistance R_S is connected in series with *A* and *B* windings. The operation also is similar to Fig. 3.23.

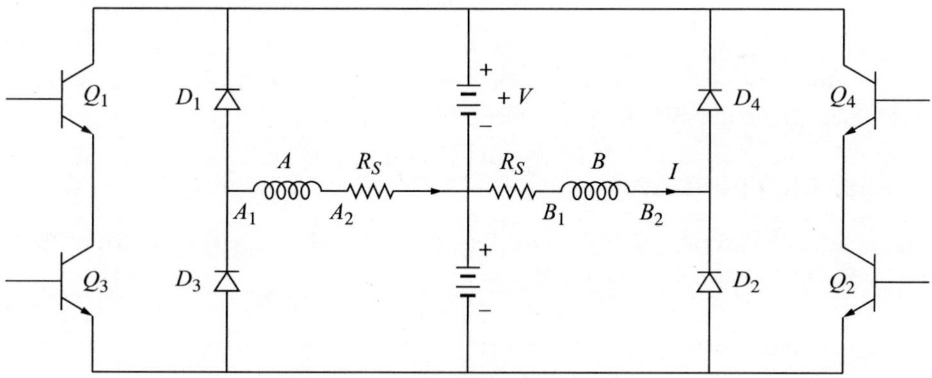

Fig. 3.23

Figure 3.24 shows a variation over figure above. Here phase winding *A* and *B* are in two sections which are connected in parallel due to this winding resistance reduced to half and current rating with be doubled. This circuit operation is also similar to Fig. 3.23.

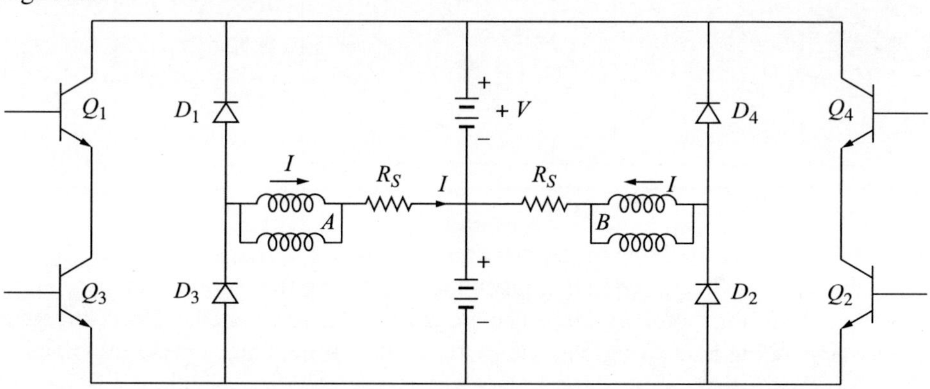

Fig. 3.24

Figure 3.25 shows bipolar L/R drive with unipolar supply. This operation is similar to to Fig. 3.22.

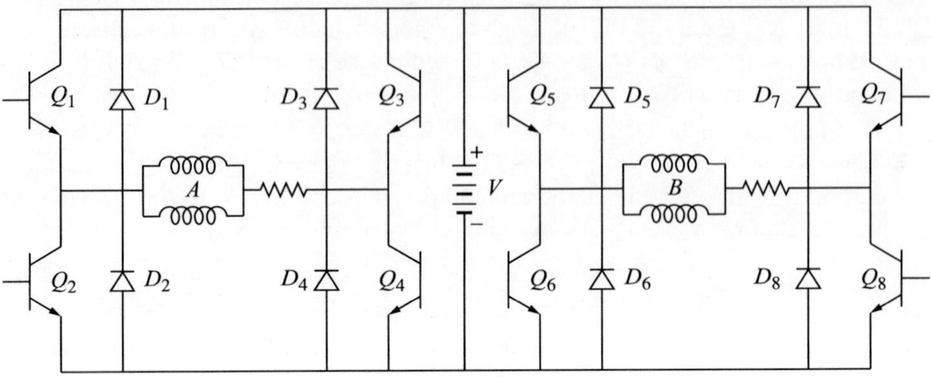

Fig. 3.25

By using L/R drives, electrical time constant of the motor winding reduces. Hence current rises rapidly, through the motor winding. Consequently much higher current and hence torque are available of higher stepping rates.

Bipolar Chopper Drives

Figure 3.26 shown bipolar chopper drives from 4 lead stepper motor. This stepper motor has two stator phase A and B.

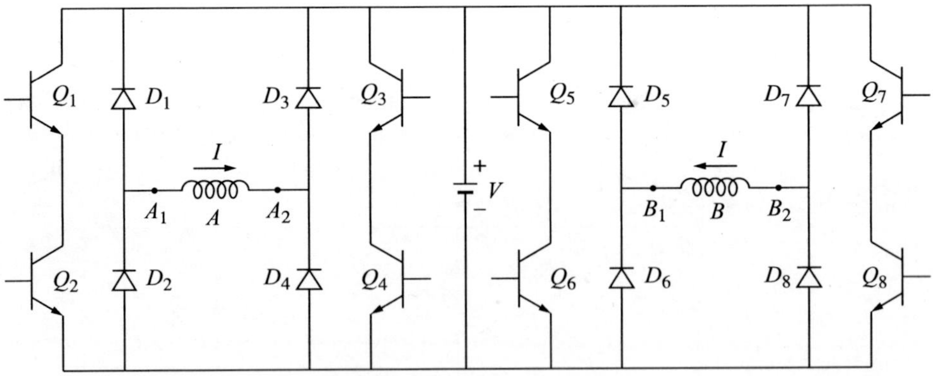

Fig. 3.26

This chopper drives requires a much higher voltage from power supply than the rated voltage of the motor winding. The motor current rise and hence torque are much higher at stepping rates. The dynamic performance of the motor is also improved.

Comparison between Bipolar and Unipolar Drives

Bipolar drive	*Unipolar drive*
1. 100% of phase a winding is utilized of winding any time.	Current flows through either A_1 or A_2 at any time. Thus only 50% of phase A winding is utilized of any time.
2. Circuit is complex and require larger number of switching transistor	Circuit is simple and require lower number of switching tranistor.
3. Motor develops large torque and more effective use of the motor winding.	Motor develops lower torque.
4. Current flow in both direction.	Current flow in one direction.

Advantages of Bipolar Drive

1. Bipolar drives lies in the utilization of copper in the motor.
2. 100% of phase A winding is utilization of any time.
3. By using bipolar drives, the stepper motor develops large torque and more effective use of the motor winding.
4. The number of phase windings in bipolar is half that of unipolar drive. Hence step angle errors due to manufacturing tolerances of the winding is also halved.

Disadvantages

1. The circuit becomes complex.
2. It requires larger number of switching transistors.
3. Due to the time constant of the motor winding is increased torque available is reduced of higher stepping rates.

3.9 SWITCHED RELUCTANCE MOTOR (SRM)

3.9.1 Introduction

The switched reluctance motor is a double salient, singly excited motor, *i.e.*, it has salient poles in both the rotor and stator, but only the stator carries windings. The rotor has no windings, magnets. It works on variable reluctance principle. Operates on the principle of Faraday's law of electromagnetic induction.

Faraday's Law

Whenever the flux linkages associated with a electric circuit, an emf is induced in it.

Flux linkage $\lambda = N\phi$

SRM are used in adjustable-speed drives because of the following features:

- Motor is simple in construction with no winding on rotor and simple concentrated coils on stator.
- It can be run at high speeds (about 2×10^5 rpm) because of no windings on rotor and rugged rotor construction.
- Stator windings can be cooled easily.

- It can be operated from unidirectional drive circuits.
- SRM operates successfully though at reduced output, even if one or more phases are out of circuit due to some fault.
- These motors can be manufactured with large number of stator and rotor teeth, they give large torque per unit volume.

Construction of SRM.

Block Diagram

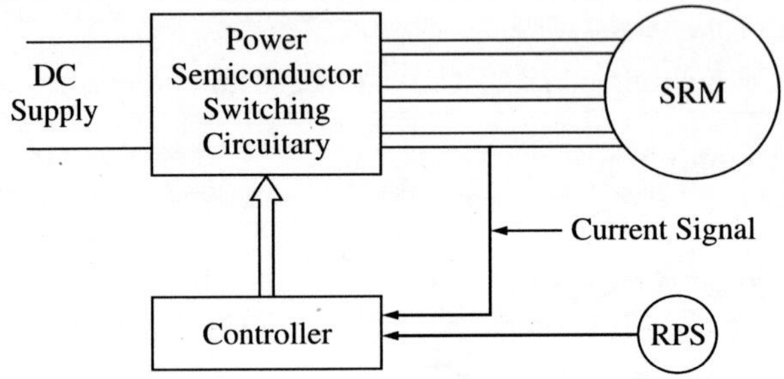

Fig. 3.27

Constructional Diagram

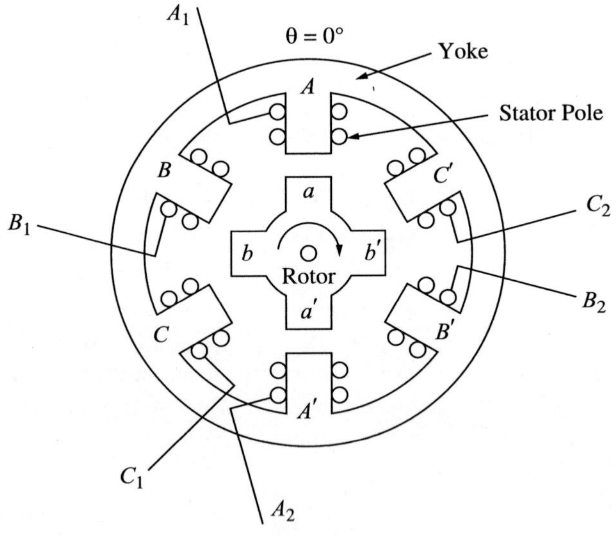

Fig. 3.28

Terminal Board

$$A_1^{\bullet}\ A_2^{\bullet}\ B_1^{\bullet}\ B_2^{\bullet}\ C_1^{\bullet}\ C_2^{\bullet}$$

Stator

- Stator is made up of silicon steel stampings with inward projected poles.
- The number of poles of the stator can be either in even number or odd number. Most of the motors have even number of stator poles.

- All these poles carry field coils (or) stator windings. The field coils of opposite poles are connected in series such that mmf's are additive and they are called phase windings. Phase windings are connected to the terminals of the motor.

Rotor

The rotor is also made up of silicon steel stampings with outward projected poles. The rotor shaft consists of rotor position sensor. Number of poles of the rotor are different from the number of stator poles.

- Figure 3.27 shows the block diagram of SRM. DC supply is given to the power semiconductor switching circuitary, which is connected to the various phase windings of SRM.
- RPS which is mounted on a shaft of the SRM, provides signal to the controller about the position of the rotor with respect to the reference axis.
- Controller collects these information and also reference speed signal and suitably turns on and off the concern power semiconductor devices of switching circuit, such that the desired phase winding is connected to the DC supply.
- The current signal is also a feedback to the controller circuit to unit the motor current within permissible limits.

3.9.2 Principle of Operation

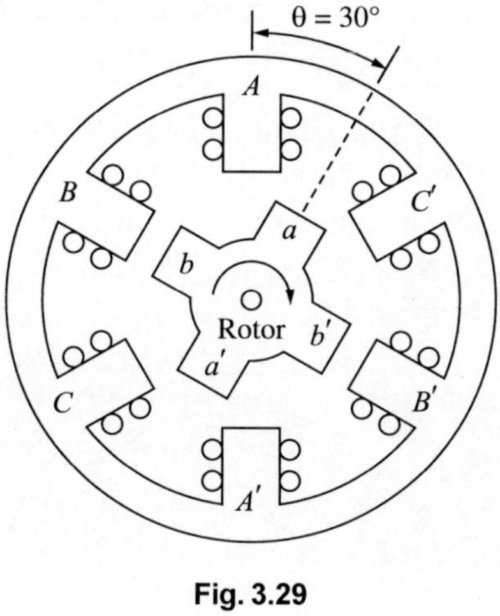

Fig. 3.29

$$\frac{\partial L_B}{\partial \theta} = 0$$

- Stator poles A_1A' and rotor poles a_1a' are in alignment, they are in minimum reluctance position so far as phase '*A*' is concerned *i.e.*, at this position inductance '*B*' phase winding is neither maximum nor minimum.

In this condition $\dfrac{\partial L_a}{\partial \theta} = 0.$

- Now there axists $\frac{\partial L_B}{\partial \theta}$. At this position, if the phase winding '*B*' is energized, then the motor develops a torque because of variable, reluctance principle because of $\frac{\partial L_B}{\partial \theta}$.
- The motor develops a torque which is called as "electromagnetic torque" which is equal to $\frac{1}{2} i_B^2 \frac{\partial L_B}{\partial \theta}$. The direction of this torque is such that B_1B' and b_1b' try to get alignment.
- If the torque is more than the opposing load and frictional torque, the rotor begins to rotate, when the shaft occupies the position such that BB' and bb' are in alignment, then no torque is developed because $\frac{\partial L_B}{\partial \theta}$ is zero.
- Now phase winding '*B*' is switched off and phase winding '*c*' is turned on to dc supply.
- The load experiences a torques as $\frac{\partial L_C}{\partial \theta}$ exists, the electromagnetic torque developed $\frac{1}{2} i_C^2 \frac{\partial L_C}{\partial \theta}$. The motor continues to rotate.
- When the rotor moves further 30° the torque due to '*C*' winding is zero. Then it is switched off and phase winding '*A*' is energized. This is a continuous and cyclic process. Thus rotor starts. It is a self starting motor.
- The speed increases, the load torque requirement also changes, when the average developed torque is more than the load torque.
- The rotor attains dynamic equilibrium condition when developed torque is equal to load torque.

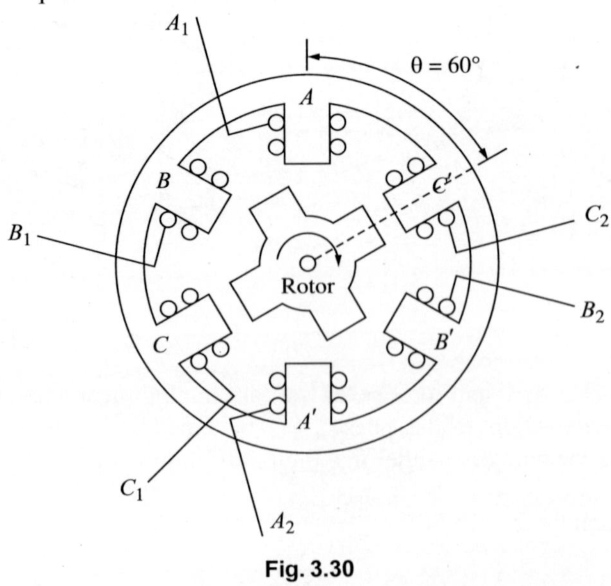

Fig. 3.30

- At this steady state condition, the power drawn from the mains is equal to time rate of change of energy stored in magnetic field.
- When the load is increased, speed of the motor tends to fall. So that power balance is maintained, then torque developed is to be increased by increasing the current. Thus more power is drawn from the mains vice versa takes place when load is reduced. Thus electrical to mechanical power conversion takes place.

3.9.3 Voltage and Torque Equation of a SRM

Basic voltage equation of SRM

$$V = iR + \frac{d\lambda}{dt} \qquad [\therefore \lambda = Li]$$

$R \rightarrow$ resistance of phase winding

$\lambda \rightarrow$ flux linkage

$$\frac{d\lambda}{dt} = \frac{d(Li)}{dt} = L\frac{\partial i}{\partial t} + i\omega\frac{\partial L}{\partial \theta}$$

$$V = iR + L\frac{\partial i}{\partial t} + i\omega\frac{\partial L}{\partial \theta}$$

where $\frac{\partial \lambda}{\partial i} \rightarrow$ slope of magnetic curve

$\partial L \rightarrow$ incremental inductance

$iR \rightarrow$ resistive drop

$L\frac{\partial i}{\partial t} \rightarrow$ emf due to incremental inductance

$i\omega\frac{\partial L}{\partial \theta} \Rightarrow e \rightarrow$ "Self emf" depends on currents speed and rate of change of inductance with rotor angle.

Thus the equivalent circuit of SRM consists of each phase, a resistance, an incremental inductance and self emf. $L\frac{\partial i}{\partial t}$ emf due to incremental inductance is zero.

During the flat-top period, emf 'e' is constant. At some instant the inductance is constant, 'e' will be zero and $L\frac{\partial i}{\partial t}$ will be constant.

Thus the equivalent circuit of SRM changes from being mainly on inductance and emf. 'X' plying Eqn. (1) by (i) on both sides

$$V_i = i^2R + Li\frac{\partial i}{\partial t} + i^2\omega\frac{\partial L}{\partial \theta} \qquad ...(2)$$

$V_i \rightarrow$ electrical energy supplied in watts.

$i^2R \rightarrow$ resistive loss.

$Li\frac{\partial i}{\partial t} \rightarrow$ Power associated with incremental inductance.

$ei \rightarrow$ Power due to self emf (*i.e.*)

(*i*) Power associated with change in stored energy.

(*ii*) Power converted into mechanical.

- Stored energy in the magnetic field $= \frac{1}{2}Li^2$
- Power associated with change in stored energy $= \frac{dW_{st}}{dt}$

$$= \frac{1}{2}L(2i)\frac{\partial i}{\partial t} + \frac{1}{2}i^2\frac{\partial L}{\partial \theta} \times \omega$$

$$= Li\frac{\partial i}{\partial t} + \frac{1}{2}i^2\,\omega\frac{\partial L}{\partial \theta} \qquad \text{...(3)}$$

Power converted into mechanical,

$P_m = Vi - i^2R$ – Power associated with change in stored energy. ...(4)

Substituting Eqn. (2) and (3) in (4), we get

$$= Vi - i^2R - Li\frac{\partial i}{\partial t} - \frac{1}{2}i^2\omega\frac{\partial L}{\partial \theta}$$

$$= i^2R - i^2R + Li\frac{\partial i}{\partial t} - Li\frac{\partial i}{\partial t} + i^2\omega\frac{\partial L}{\partial \theta} - \frac{1}{2}i^2\omega\frac{\partial L}{\partial \theta}$$

$$\therefore \quad P_m = \frac{1}{2}i^2\omega\frac{\partial L}{\partial \theta}$$

$$P_m = \omega T$$

Troque developed by an SRM is

$$T = \frac{1}{2}i^2\frac{\partial L}{\partial \theta} \text{ N-m}$$

∴ Applications of SRM

- Washing machine
- Vacuum cleaners
- Fans
- Future automobile applications
- Rototics control applications.

3.9.4 Operating Modes

These are two types of operating modes of SRM:

1. Single-pulse mode
2. Pulse-width modulation (PWM) mode.

1. Single Pulse Mode

In single pulse mode, also called high speed mode, the current rises in within limits during small interval of each phase excitation. This built up current limit is due to winding inductance and motional counter emf generated in stator winding. An ideal current waveform and variation of L with θ is shown in Fig. 3.31(a).

When $L(\theta)$ begins to 0, switch is closed and current at once reaches a const. value. When $L(\theta)$ stops rising, switch is opened and current decays to 0 immediately. During the time current pulse is const. and +ve, $L(\theta)$ is rising and therefore +ve torque is produced. Actually ideal waveforms are non-existant.

In Fig. 3.31(b) V_S should be applied to winding some what in advance of instant when $L(\theta)$ begins to rise. Low L allows $i(\theta)$ to build up faster to large value before torque producing region begins.

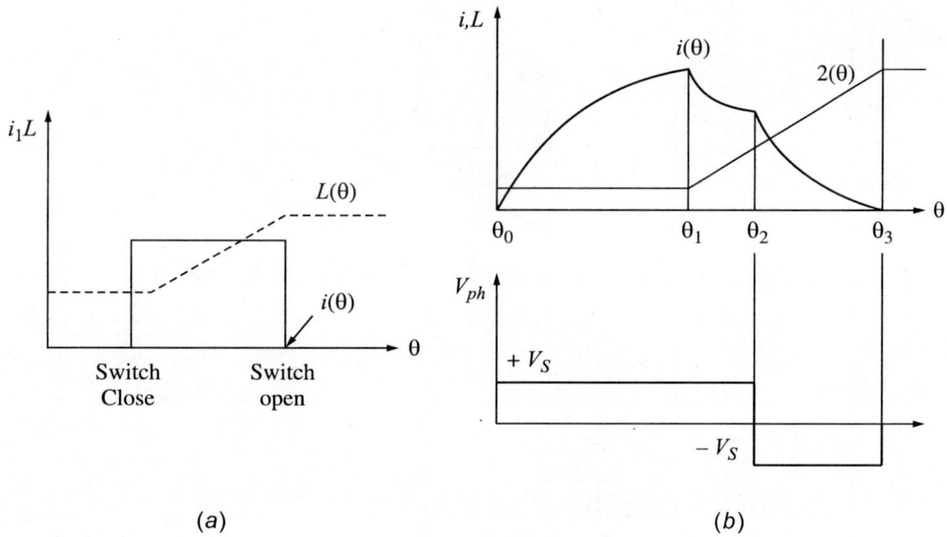

Fig. 3.31

After θ_1, current $i(\theta)$ begins to fall of rising $L(\theta)$ and motional emf switch is opened at θ_2. At this rotor angle θ_2, –ve voltage is applied to phase winding. This cause rapid fall of current and $i(\theta)$ finally reduces to 0 at extinction angle θ_3. Angle $(\theta_2 - \theta_3)$ is transistor conduction angle and $(\theta_3 - \theta_2)$ is diode conduction angle of an inverter.

Source delivers energy to motor during $(\theta_2 - \theta_0)$ and energy is feedback by motor to source during $(\theta_3 - \theta_2)$. The angle $(\theta_3 - \theta_0)$, called θ_c (conduction angles), is an important control parameter. θ_0 is called switch *on* angle and θ_2 is switch *off* angle.

$(\theta_1 - \theta_0)$ is called angle of advance.

The –ve voltage to phase winding at switch off θ_2 should be applied much before the max. value of $L(\theta)$ is reduced. This is done to ensure that $i(\theta)$ decays to '0' and –ve torque region is avoided.

2. PWM Mode: (For Chopping)

In this mode, also called low-speed mode, each phase winding gets excited for a period sufficiently long. In order to keep current rise within acceptable ratings of motor and inverter components, a current limiting device is incorporated before SRM.

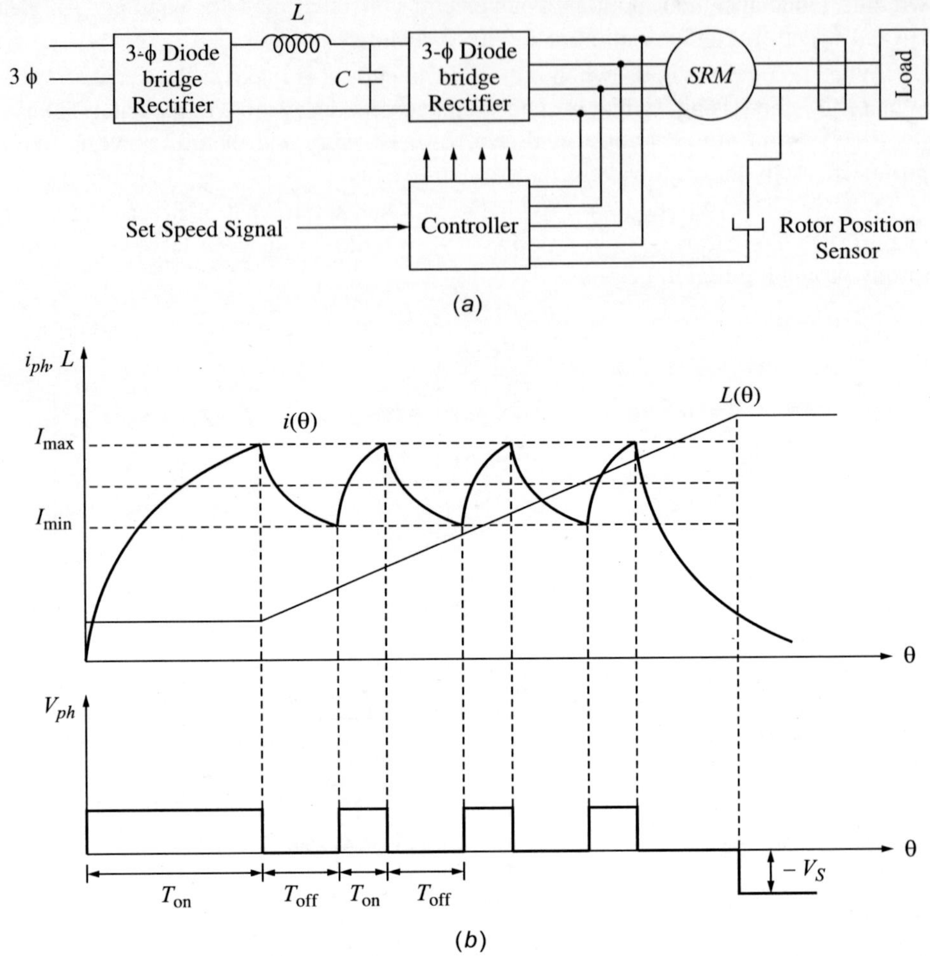

Fig. 3.32

This is achieved by installing a current sensor in each phase so as to monitor current. The current sensor then controls alternately the *on* and *off* instants of power converter components in order to hold the current between permissible upper and lower levels. The chopped phase current as a function of rotor position θ is shown in Fig. 3.32(*b*).

In Fig. 3.32, I_{max} and I_{min} are upper and lower prescribed levels of chopped phase current and I_0 is its mean value. The λ variation is shown in Fig. 3.32(*b*).

If upper and lower limits of each phase current are shifted up or down, then mean value of motor current would alter. This will results in torque control and hence speed adjustment as desired.

Merits

1. Construction is simple and robust.
2. Rotor carries no windings, no brushes, less maintenance.
3. There is no permanent magnets.

4. No shoot through fault.
5. Switching circuitry is simple.
6. It is self-starting machine.
7. Possibility to get very high speed.
8. Operation of machine can be easily change from motoring mode to generating mode.

Demerits

1. Stator phase winding should be capable of carrying magnetising current also.
2. It requires a rotor position sensor.
3. No. of power wires between the power semiconductor circuitry and motor and no. of control cables from controller to power semiconductor switching circuitry are more.
4. For high speed current waveform has undesirable harmonics.
5. For high speed operation developed torque has undesirable ripples which produces noise.

3.10 POWER SEMICONDUCTOR SWITCHING CIRCUITS FOR SRM (POWER CONTROLLER SWITCHING CIRCUITS)

The selection of a converter depends upon the application. One of the main aspects of the research in SRM drives has been the converter design. The main objectives of the design of the converter are performance of the drive and cost of the drive.

Basic Requirements

(*i*) Each phase of SRM should be able to conduct independent of the other phases.

(*ii*) The converter should be able to demagnetize the phase before it steps into the generating region if the machine is operating as a motor and should be able to excite the phase before it steps into the generating region if operated as a generator.

(*iii*) The converter should be able to freewheel during the chopping period to reduce the switching frequency.

(*iv*) The converter should be able to utilize the demagnetization energy from the off going phase in a useful way by either feeding it back to the source or using it in the next conducting phase.

3.10.1 Clamic Converter

The most versatile SRM converter topologies the clamin bridge converter with two power switches and two diodes perphase shown in Fig. 3.33.

During the conduction mode of phase A, T_1 and T_2 are in on state. This mode is usually initiated before the start of the rotor and stator pole overlap. So that the phase covert reaches the reference value, before the phase inductance begins to increase this helps to reduce the torque ripple.

In this mode the current is maintained at the reference value by switching one of the phase switches while leaving the other one continuously on till the commutation time is reached.

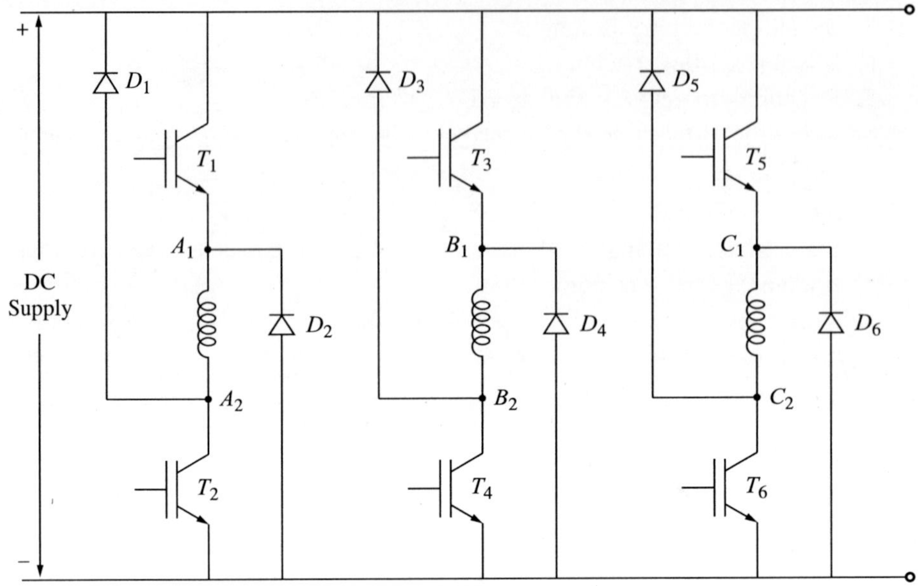

Fig. 3.33

Both of the phase switches are turned off to initiate commutation. The phase starts, to demagnetize through the two diode and the energy transfers from the motor phase to the dc source.

During the commutation the off going phase winding for a voltage of $-V_{dc}$.

3.10.1.1 Operation

Phase winding A in connected to dc supply through power semiconductor switches T_1 and T_2, depending upon the rotor position. When phase winding A is to be energized the device T_1 and T_2 are turned on. When T_1 and T_2 are turned off, the phase winding A is to be disconnected from the supply. But stored energy in the phase winding A tends to maintain the current in the same direction. This current passes through diode D_1 and D_2 to the supply. Thus stored energy is feedback to the mains, then phase windings B and C are switched on and off from the supply in a cycle manner.

Merits

(*i*) Control of each phase is completely independent of the other phases.

(*ii*) The converter is able to Freewheel during the chopping period of low speeds, which helps to reduce the switching frequency and then the switching laws of the converter.

(*iii*) The energy from the off going phase is feedback to the source which results in useful utilization of the energy.

Demerits

High number of switches required in each phase which makes the converter expensive and also used for low voltage application.

3.10.2 (n + 1) Power Semiconductor Switches and (n + 1) Diodes per Phase, where 'n' is Number of Phase Windings

Figure 3.34 shows an $(n + 1)$ power semiconductor switches and $(n + 1)$ diodes per phase where 'n' is number of phase windings.

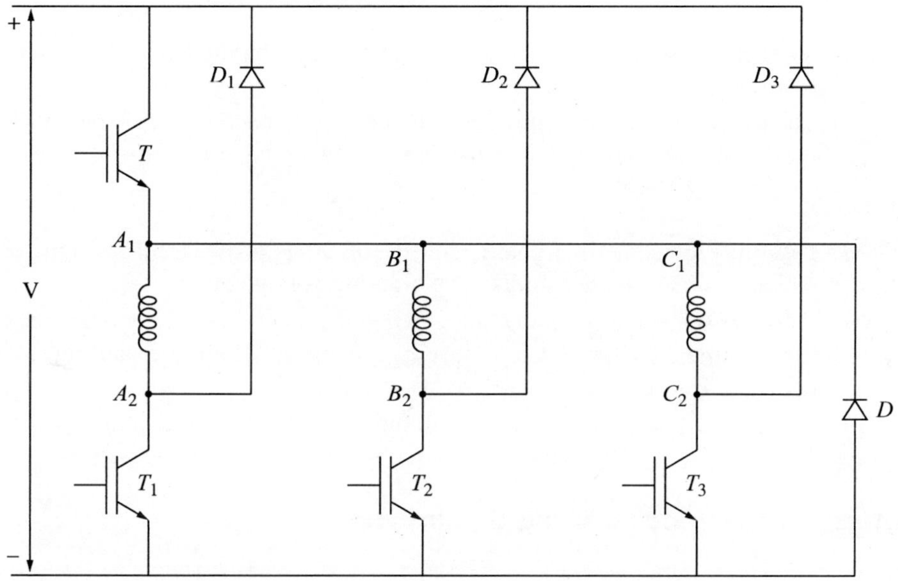

Fig. 3.34

The commutation is initiated by turning both the common switch and the phase switch off.

The phase demagnetizes through the common diode and the phase diode and the energy is transferred back to the source. During the commutation, the phase winding sees a voltage of $-V_{dc}$.

The next phase cannot be turned on until the off going phase has not been completey demagnetized.

In order to start the magnetization of the next phase common switch has to be turned on.

This gives a path to the demagnetization currents of the off-going phase forcing zero voltage across it and therefore allowing the demagnetization process.

3.10.2.1 Operation of the Circuit

It 3ϕ motor requires 4 transistors our diodes when IGBTs T and T_1 are turned on. Phase winding 'A' is energized then phase winding 'A' connected to the dc supply.

When IGBTs T and T_1 are turned off phase winding 'A' is disconnected from the supply. But energy stored in the winding in fedback to the mains through diodes D and D_1.

When IGBTs T and T_2 over turned on, the phase winding B is connected to the mains. T and T_2 are turned off. Stored energy in phase winding B is fedback to dc mains through diodes D and D_2.

Similarly phase winding 'C' is energized the cycle is repeated.

Merits

(*i*) The converter uses lower number of switching devices which reduces the cost of the converter.

(*ii*) The converter is able to freewheel during the chopping then reducing the switching frequency and losses.

(*iii*) Voltage rating of all the switching devices and the diodes are V_{dc} which is relatively low.

(*iv*) The energy for the off going phase is transferred back in to the source which results in useful utilization of the energy and also improves the efficiency.

Demerits

(*i*) Disability to magnetize a phase while the off going phase is still demagnetizing which results in higher torque ripple during commutation.

(*ii*) A higher speed the off going phase cannot be -energized fast enough because the common switch *T* keeps turning on intermediately disabling forced demagnetization.

(*iii*) The common switch conducts for all the phase and thus, has higher switching stress.

3.10.3 Phase Winding using Bipolar Wires

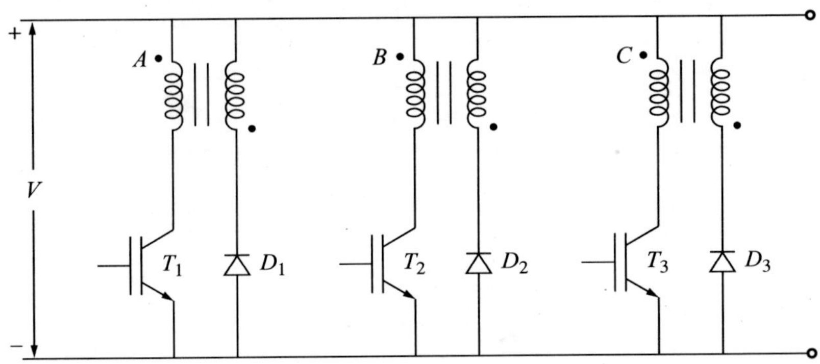

Fig. 3.35

3.10.4 Split Power Supply Converter

Operation: Figure 3.36 shows the circuit of split power supply converts the main power supply $2V_{dc}$ is split into two helver using split capacitors. During conduction energy in supplied to the phases by one half the power supply. During commutation period the phases demagnetize into other half of the power supply.

When IGBT T_1 is turned on the phase winding 1 is energized by capacitor C_1 when IGBT T_1 is turned off. Stored energy is the phase winding 1 in fedback to the capacitor C_2 through diode D_2.

When IGBT T_2 in turned on by capacitor C_2 and phase winding 2 is energized when IGBT T_2 is turned off. Stored energy in the widing 2 is fedback to the capacitor C_1 through diode D_1. The similar operation taken place in the remaining winding also.

3.10.4.1 Operation

Figure 3.36 shows the above switching circuit of phase winding using bipolar wires. Each stator pole carriers a coil using bipolar wire. Each phase group has two exactly identical phase windings and they have a common magnetic case T_1 is turned on, phase winding A is energised winding is connected to the dc mains.

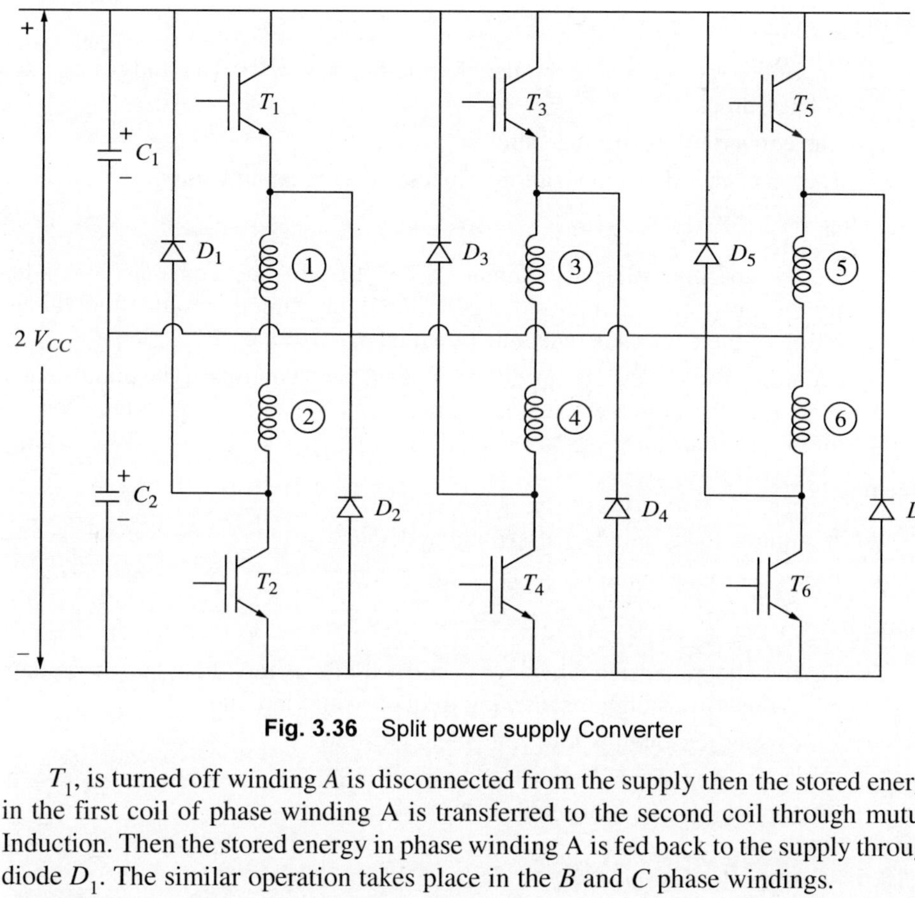

Fig. 3.36 Split power supply Converter

T_1, is turned off winding A is disconnected from the supply then the stored energy in the first coil of phase winding A is transferred to the second coil through mutual Induction. Then the stored energy in phase winding A is fed back to the supply through diode D_1. The similar operation takes place in the B and C phase windings.

Merits:

(*i*) The converter used lower number of switching devices thus reducing the cash on the converter.

(*ii*) The converter allows last demagnetization of phases during commutation.

Demerits:

(*i*) Free wheding in not possible during chopping as the phases. See the $-V_{dc}$ this causes a higher ripples in current and torque during chopping.

(*ii*) The implication in the coupling between the two winding causes voltage spikes during turn-off.

(*iii*) The clipper lower anociated with the auxiliary winding are unacceptably high for many applications.

During the commulation period, D_1 is forward biased and the energy from the machine phase is transferred to the dump capacitor C_d.

The excess energy from the dump capacitor 'C_d' is transferred into the source by turning on the switch θ_d when the switch θ_d is off. Energy is transferred to the source through 'D_c'.

Merits:

(*i*) This topology uses lower number of switching devices and has only one switch voltage drop.

(*ii*) The converter has full regenerative capability.

(*iii*) There is faster demagnetization of phases during commutation.

Demerits:

(*i*) Dump capacitor voltage is maintained $2V_{dc}$ to allow fast demagnetization but use of a capacitor and an inductor in the dump circuit and also the voltage rating of the devices is twice the by voltage.

(*ii*) Monitoring of the dump capacitor voltage C_d and controlar. The dump switch θ_d makes the converter very complicated and also the converter does not allow free wheeling.

Merits:

(*i*) It requires lower number of switching devices.

(*ii*) Faster demagnetization at phases during commutation.

Demerits:

(*i*) During chopping, free wheeling is not possible as the phase see the voltage $V_{dc}/2$ this causes higher switching frequency and more lesses.

(*ii*) This is not feasible for low voltage application.

(*iii*) The converter is less fault tolerant and fault in any phase will unbalance the other phase that in connected to it.

3.10.5 Dump 'C' Converter

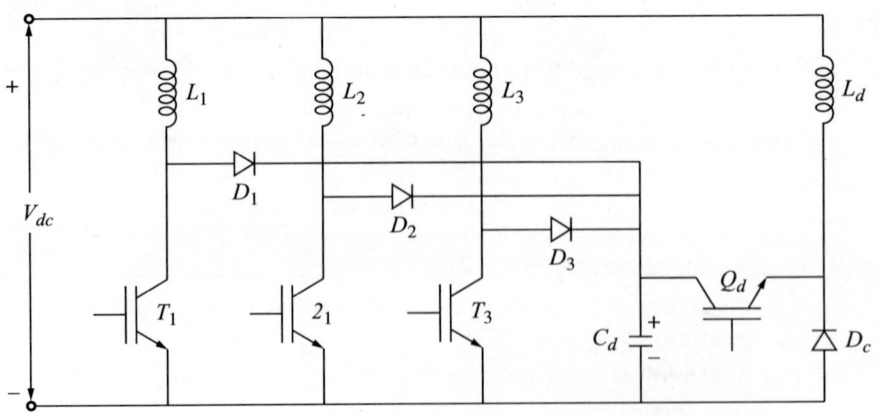

Fig. 3.37 Basic 'C' dump converter circuit

Figure 3.37 above shows the dump 'C' converter switching circuit for SRM. The energy from the off going phase in dumped into a capacitor to achieve fast demagnetization. This energy is then returned to the source from the capacitor.

The conduction of the phase initiated by turning on the phase switch T_1 for 1 connected in series with the phase. The phases are demagnetized by turned off the phase switch T.

SOLVED NUMERICALS

Example 3.1

A stepper motor driver by a bipolar drive an unit has the following parameters. Winding inductance = 30 mH, rated current = 3A, DC supply = 45 V, total resistance is each phase = 15 Ω. When the transistors are turned off determine.

(i) the time taken by the phase current to decay to zero.

(ii) the proportion of the stored inductive energy returned to the supply.

Solution:

(*i*) The current in the windings decay to zero in time.

$$t' = 1/2\ \tau_{ed}$$

$$\tau_{ed} = \frac{Lm}{12m} = \frac{30 \times 10^{-3}}{15} = 2ms$$

$$t' = \frac{2 \times 10^{-3}}{2} = 1ms$$

(*ii*) The energy returned to supply during current decay.

$$1/2\ L_m\ I_{ph}^2 = \frac{1}{4} \times 30 \times 10^{-3} \times 3^2$$

$$= 67.5 \text{ mW}$$

Example 3.2

What is the step angle of a four phase stepper motor with 12 stator teeth and 3 rotor teeth?

Solution: Step Angle $\beta = \dfrac{360°}{mN_r}$

Here

$$n = 4$$

$$N_s = 12$$

$$N_r = 3$$

$$\beta = \frac{360°}{4 \times 3} = 30°$$

$$\beta = 30°$$

Example 3.3

A hybrid 4R stepping motor has 10 main poles which have been castellated to have 7 teeth each. If rotor has 60 teeth. Calculate the step angle.

Solution: No. of station teeth

$$N_s = 10 \times 7 = 70$$

No. of rotor teeth

$$N_r = 60$$

Step angle

$$\beta = \frac{N_s - N_r}{N_s \cdot N_r} \times 360°$$

$$= \frac{70 - 60}{70 \times 60} \times 360°$$

$$= 0.857°$$

Example 3.4

A stepper motor has a step angle of 3°. Determine (a) resolution (b) number of steps required for the shaft to make 25 revolutions and (c) shaft speed if the stepping frequency is 3600 pps.

(*a*) Resolution $= \dfrac{360°}{\beta} = \dfrac{360}{3} = 120$ step/res.

(*b*) Now, steps/revolution = 120.

Hence steps required for making 25 revolutions = 120 × 25 = 3000.

(*c*) Speed $n = \dfrac{Bf}{360°} = \dfrac{3 \times 3600}{360°} = 30$ nps.

Example 3.5

A stepper motor has a step angle of 1.8° and is driven 4000 pps. Determine (i) resolution (ii) motor speed (iii) Number of poles required to rotate the shaft through 54°.

Solution:

(*i*) Resolution $= \dfrac{360°}{\beta} = \dfrac{360}{1.8} = 200$ steps/rev.

(*ii*) Motor speed $n = \dfrac{\beta \times f}{360°} = \dfrac{1.8 \times 4000}{360°} = 20$ rps

(*iii*) Number of pulses required to rotate the shaft through 54°

$$= \frac{54}{1.8} = 30 \text{ pulses.}$$

Example 3.6

Calculate the step L of a single stack, 4 phase, 8/6-pole 1/R stepper motor. What is its resolution.

Solution: Step angle
$$\beta = \frac{N_s - N_r}{N_s \cdot N_r} \times 360°$$
$$= \frac{8-6}{8 \times 6} \times 360° = 15°$$
$$\text{Resolution} = \frac{360°}{\beta} = \frac{360}{15} = 24 \text{ steps/revolution}$$

Example 3.7

A single-stack, 3 phase VR motor has a step of –1.5°. Find the number of its rotor and stator poles.

Solution:
$$\beta = \frac{360°}{mN_r}$$
$$M_r = \frac{360°}{m\beta}$$
$$= \frac{360°}{3 \times 15} = 8$$
$$M_r = 8$$

for finding the value of N_s, we use the relation.

$$\beta = \frac{N_s \sim N_r}{N_s \cdot N_r} \times 360°$$

(*i*) When $N_s > N_r$.

$$\beta = \frac{N_s \sim N_r}{N_s \cdot N_r} \times 360°$$
$$15° = \frac{N_s \sim 8}{N_s \cdot 8} \times 360°$$
$$N_s = 12$$

(*ii*) When $N_s < N_r$

$$15 = \frac{N_r - N_s}{N_s . 8} \times 360°$$
$$= \frac{8 - N_s}{8 \times N_s} \times 360°$$
$$N_s = 6.$$

Example 3.8

A four-stack VR stepper motor has a step angle of 1.5°. Find the number of its rotor and stator teeth.

Solution: A four stack motor has four phases $n = 4$

$$\beta = \frac{360°}{MN_r}$$

$$N_r = \frac{360°}{m\beta} = \frac{360°}{4\times 1.5} = 60$$

Since, it is multi stack motors, rotor teeth equal to the stator teeth *i.e.*, $N_s = 60°$.

Example 3.9

Calculate the pulse rate required to obtain a rotor speed of 2400 rpm for a stepper motor having a resolution of 250 steps/rev.

Solution: Motor speed = 2400 rpm.

$$n = \frac{2400}{60} = 40 \text{ rps}$$

$$\text{Step angle } \beta = \frac{360}{\text{No. of steps/rev}}$$

$$= \frac{360}{200} = 1.8°$$

$$n = \frac{\beta\times f}{360°}$$

$$f = \frac{n\times 360°}{\beta} = \frac{40\times 360°}{1.8} = 8000 \text{ pps}$$

$$f = 8000 \text{ pps}$$

Example 3.10

A stepper motor has a resolution of 500 steps per on the 1-phase on mode of operation. If it is operated in half step mode, determine (a) Resolution (b) number of steps required to turn to rotor through 72°.

Solution: For full step mode, the steps/rev. = 500

$$\text{Now half step mode} = \frac{500}{2} = 1000 \text{ steps per}$$

$$\text{Step angle } \beta = \frac{360°}{1000} = 0.36°$$

$$\text{No. of steps required to turn the rotor through } 72° = \frac{72°}{0.36} = 200 \text{ steps}$$

Example 3.11

A stepper motor has a step angle of 1.8°. What number should be loaded into the encoder of its drive system if it is desired to turn the shalt ten complete revolutions?

Solution: Step angle

$$\beta = 1.8°$$

$$\text{No. of steps/rev} = \frac{360°}{1.8} = 200 \text{ steps/rev.}$$

$$\text{No. of step/10 rev} = 200 \times 10 = 2000 \text{ steps.}$$

Example 3.12

What is the motor torque T_m required to accelerate an initial load of $10^{-4}/gm^2$ from w_1 = 200 to w_2 = 300 rad/sec during 0.2 sec Frictional load torque in 0.06 N – m.

Given data

$$J = 10^{-4}\ kgm^2,\ w_1 = 200\ rad/sec$$

$$w_2 = 300\ rad/sec,\ \Delta f = 0.2\ sec.$$

$$T_f = 0.06\ N-m$$

To find T_m

Solution: $\frac{dw}{df} = \frac{w_2 - w_1}{\Delta f} = \frac{300 - 200}{0.2} = 500$

$$\text{Motor torque } T_m = J\frac{dw}{dt} + T_f = 10^{-4}(500) + 0.06$$

$$T_m = 0.11\ N\text{–}m$$

Example 3.13

What is the motor torque T_m required to accelerate an initial load of 3 × 10^{-4} kgm^2 from f_1 = 1000 Hz to f_2 = 2000 M_2 during 100 msec. The frictional torques T_f in is 0.05 N-m and step angle in 1.8°.

Given data

$$J = 3 \times 10^{-4}\ kgm^2 \quad f_1 = 1000\ Hz \quad f_2 = 2000\ Hz.$$

$$\Delta t = 100\ m\ sec \quad T_f = 6.05\ N.M \quad Q_s = 1.8°$$

To find *Motor torque (T_m)*

Solution: Step angle Q_s. In reactions = Q_s in degrees × $\frac{\pi}{180}$

$$= 1.8° \times \frac{\pi}{180} = 0.0314 \text{ rad}$$

$$w_1 = Q_5 f_1 = 0.0314 \times 1000 = 31.4 \text{ rad/sec}$$

$$\frac{dw}{dt} = \frac{w_2 - w_1}{\Delta t} = \frac{62.8 - 31.4}{100 \times 10^{-3}} = 314 \text{ rad/sec.}$$

$$\text{Motor torque } T_m = J\frac{dw}{df} + T_f = 3\times10^{-4} + 314 + 0.05$$

$$T_m = 0.1442 \text{ N-m}$$

Example 3.14

What is the maximum acceleration of an initial load of 4 × 10^{-4} kgm^2 driven by a motor torque of 0.3 Nm. Frictional loaded are negligible and step angle in 3°.

Given data: $J = 4 \times 10^{-4}$ kgm^2, $Jm = 0.3$ N – M

$$Q_s = 3° = 3x\frac{\pi}{180} = 0.0523 \text{ rad/sec.}$$

To find

Maximum acceleration

Solution:

$$T_m = JQ_S\frac{df}{dt}$$

$$\frac{df}{dt} = \frac{T_m}{JQ_S} = \frac{0.3}{4\times10^{-4}\times0.0523} = 1.434\times10^4 \text{ step/sec}^2$$

Maximum acceleration = 1.434 × 10^4 steps/sec^2

SOLVED PROBLEMS

Example 3.1

What is the step angle of a 3ϕ SRM having 12 stater poles and 8 rotor poles what is commutation frequency at each phase of the speed of 6000 rpm.

Given data

$$N_s = 12 \text{ poles} \qquad q = 3, \quad N_r = 8 \text{ poles.}$$

Solution:

(*i*) Step angle = $\frac{2\pi}{qN_r} = \frac{360°}{3\times P} = 15°$

(*ii*) Commutation frequency at each phase = $\frac{N_r\times w}{2\pi} = \frac{8\times6000}{60} = 800$ Hz

Example 3.2

What is the step angle of an 5 phase SRM and commutation frequency in each phase for the speed of 600 rpm. SRM having 10 stator poles and 4 rotor poles.

Solution: $\frac{2\pi}{qN_r} = \frac{360°}{5\times4} = \frac{360°}{20} = 18°$

Commutation frequency of each phase = $\frac{N_r\times w}{2\pi} = \frac{4\times600}{60} = 400$ Hz

Example 3.3

A SRM with 6 stator poles and 4 rotor poles has a stator pole are of 3θ degree and rotor poles are in 32 degree. The aligned inductance in 10.7 mH and unaligned inductance in 1.5 mH. Saturation can be neglected. Calculate the instantaneous torque when the rotor is 30 degree before the aligned position and the phase current is 7A. What is the maximum energy conversion one state if the current in limited to 7A. Determine the average torque corresponding to this energy conversion.

Solution: $N_s = 6, N_r = 4, \beta_s = 30°, \beta_r = 32°$

$$L_a = 10.7 \text{ mH, LU} = 1.5 \text{ mH, i} = 7\text{A}$$

$$\theta_k = \theta_1 = \frac{2\pi}{4} - \left(\frac{\beta_r + \beta_s}{2}\right) = \frac{2\pi}{4} - \frac{32+30}{2} = 59°$$

$$\theta_y = \theta_2 = \frac{2\pi}{2} - \left(\frac{\beta_r - \beta_r}{2}\right) = \frac{2\lambda}{2} - \left(\frac{32-30}{2}\right) = 89°$$

$$d\theta = \theta_2 - \theta_1 = 89 - 59 = 30° \text{ or } \frac{\pi}{6}$$

$$dL = 10.7 \times 10^{-3} - 1.5 \times 10^{-3} = 9.2 \times 10^{-3} \text{ H}$$

$$T = \frac{1}{2} i^2 \frac{dL}{d\theta} = \frac{1}{2} \times 7^2 \times \frac{9.2 \times 10^{-3}}{\pi/6}$$

$$T = 0.43 \text{ N-m}$$

$$\lambda_u = La\ xi = 10.7 \times 10^{-3} \times 7 = 0.0749$$

$$\lambda_u = Lu\ xi = 1.5 \times 10^{-3} \times 7 = 0.0105$$

W_m = mechanical energy transferred per stroke

$$= \frac{1}{2}(\lambda_a - \lambda_u) \times i = \frac{1}{2}(0.0749 - 0.0105) \times 7$$

$$= 0.2254 \text{ Joules}$$

Mechanical energy transferred per stroke revolution = $W_m \times N_r \times q$

Power transferred = Energy 1 sec.

$$P_m = 0.2254 \times N_r \times q \times \frac{N}{60}$$

$$\frac{2\pi NT}{60} = 0.2254 \times N_r \times q \times \frac{N}{60}$$

$$q = 3 \text{ phases}$$

$$N_r = 4.$$

$$T = \frac{0.2254 \times 4 \times 3}{2\pi}$$

$$T = 0.43 \text{ N-M}$$

Example 3.4

What is the flow linkage the aligned partition La = 10.7 mH, unaligned partition La = 1.5 mH. When the plane current is 7A? If the flow linkage is maintained constant while the rotor rotates from the unaligned to aligned portion of low speed. Determine the energy conversion per shake and the average torque (Given $N_r = 4$).

Solution: La = 10.7 mH, Lu = 1.5 mH, i = 7A, $N_r = 4$

$$\text{Flux linkage } \lambda_a = La \times i$$

$$= 10.7 \times 10^{-3} \times 7 = 0.0749 \text{ Wb/T}$$

$$\lambda_a = \lambda_u$$

$$0.0749 = L_u \times i_2$$

$$= 1.5 \times 10^{-5} \times i_2$$

$$i_2 = \frac{0.0749}{1.5 \times 10^{-3}} = 49.93 \text{ A}$$

Energy conversion per stroke

$$W_m = \frac{1}{2}(i_2 - i) \times \lambda$$

$$= \frac{1}{2}(49.93 - 7) \times 0.0749$$

$$W_m = 1.607 \text{ Joules}$$

$$\text{Average Torque } T = \frac{W_m q N_r}{2\pi}$$

$$= \frac{1.607 \times 3 \times 4}{2\lambda}$$

$$T = 3.07 \text{ N-m}$$

REVIEW QUESTIONS

1. What is a Stepper motor?
2. Define step angle.
3. Define slewing
4. Write down the formula for motor speed of stepper motor.
5. Define resolution.
6. What are the application of stepper motor?
7. What are the advantage and disadvantages of stepper motors?
8. What are the types of stepper motors?
9. Sketch the diagram of a *VR* stepper motor.

10. What are the different modes of excitation used in variable reluctance stepping motors?

11. Explain the construction and various modes of excitation of *VR* stepping motor?

12. What are the advantage and disadvantage of *VR* Stepper motor?

13. Explain the various modes of excitation in stepping motors.

14. Explain the construction and various modes of excitation of PM stepper motor.

15. What are the advantages and disadvantage of PM stepper motor?

16. What is hydrid stepper motor?

17. Explain the construction and working principle of hydrid stepper motor.

18. What is meant by full-step and half step operation?

19. What are advantages and disadvantages of hybrid stepper motor?

20. Differentiate between VR, PM, and hybrid stepper motors?

21. What is meant by microstepping in stepper motors?

22. Write detail about mini step drive.

23. Describe the working of a 3 stock stepper motor having 12 pole in stator and the rotor.

24. Explain the mechanism of torque production in VR stepper motors.

25. Stoke and explain static and dynamic characteristics of a stepper motor.

26. Draw the typical dynamic characteristics of stepper motor.

27. Describe the dynamic characteristics of a VR stepper motor.

28. Draw the static characteristics of stepper motor.

29. What is multi-stack VR stepper motor?

30. Define holding torque.

31. Define detente torque.

32. Define torque constant.

33. Define full an torque and full out torque.

34. What is response range?

35. What is slowing range?

36. What is synchronism in stepper motor?

37. Draw the block diagram of the drive system of a stepper motor.

38. Draw the bi-directional two phase logic sequence for four phase stepper motor and draw the truth table to explain its operation.

39. What is switched reluctance motor?

40. What are the essential differences between a stepper motor and SRM?

41. Mention any 4 advantage of switched reluctance motor.

42. Write about the disadvantage of SRM.

43. Mention some application of SRM.

44. Explain the construction and working principle of SRM.

45. Draw the simple block diagram of SRM.

46. Explain why rotor position sensor in essential for the operation of SRM.
47. What are the different types of power control of SRM.
48. Draw a schematic diagram and explain the operation of a *C* dump converter used for the control of SRM.
49. Discuss the various converter topologies for a 3 phase SR motor with merits and demerits of each. Explain any two of them.
50. Draw and explain four converter topologies for a 3-phase SRM unit the merits and demerits of each.
51. Derive the torque equation of SRM.
52. Draw and explain, torque-speed characteristics of SRM.

❑❑❑

Chapter 4

Permanent Magnet Generators

A permanent magnet generator is basically a power producer that makes use of magnets to instead of gas or any substitute power source. Since these generators make use of magnets, they do not require any maintenance.

In general, a regular generator works by burning fuel to rotate a turbine which then generates energy where as, the greatest thing about a permanent generator is that instead of using power and fuel, the magnet generator works by using the resistive and attractive forces of magnets to generate electrical energy. As a permanent magnet generator is self driven, there is no necessity to use another fuel source for it.

Advantages

- The cost incurred to create a permanent magnet generator is very low. For reason that it does not consume any burning fuel or draws power from the substitute source.
- It involves only a small amount of room space.
- This device is capable of working in all conditions *i.e.*, it works in severely hot or cold conditions without any problem.

Applications

- Wind turbines.
- Hydroelectric plant generators and other generating units.
- Paper machine main and auxiliary drives.
- Fans and screw compressors.
- Roller tables and conveyors.
- Other special system.

Typical Operating Range of PMG:

Speed : 0 – 2000 rpm

Power : upto 6 – 10 MW

4.1 INTRODUCTION TO PERMANENT MAGNET GENERATORS

The permanent magnet generators can also be called as an 'alternator', because it generates alternating current (AC). It does not generate 'mains voltage' or utility power AC. It generates as low voltage, three phase AC, then changes it into direct current (DC) for charging a 12 V battery.

The PMG consists of:

- A steel spine and shaft
- A stator containing coils of wire
- Two magnet rotors
- A rectifier

The stator consists of six coils of copper wire, cast in fibre glass resin. This stator casting is mounted on to the spine, it does not enough wires from the coils takes the electricity to the rectifier, which changes AC to DC for charging the battery. The rectifier is mounted on the aluminium 'heat sink' to keep it cool.

The magnet rotors are mounted on bearing which turn on the shaft. The rear rotor is behind the stator and enclosed with in it. The front one is on the outside, fixed to the rear one by long studs which pass through a hole in the stator. The wind turbine rotor will be mounted on the same studs. They will turn the magnet rotors, and move the magnets past the coils. Magnetic flux passes from one rotor to the other through the stator. This moving magnetic flux produces the electric power.

4.2 BRUSHLESS DC MOTORS

The term brushless dc motor is applied to many configurations of ac synchronous motors in which semiconductor control is used to control stator current such that maximum torque is obtained at a given speed. In a conventional motor the mechanical contactor, the commutator, maintains 90° electrical degrees space displacement between the rotor and stator magnetic fields to provide for the required torque.

Theoretically, the stator and rotor functions of a machine can be inverted, putting the field system on the rotor. There is no advantage to be gained if conventional commutation is used, as the commutator sections are fixed and brush gear must rotate at the speed of rotor field. Solid-state switching by transistors or thyristors, triggered by position sensors can, however, replace the brush gear by fully electronic commutation, endowing small machines with a valuable control facility.

In this method each of the stator phase winding is energized sequentially by a power transistor (or) thyristor by means of a signal from position sensor placed on the rotor. Because of the rotor position feedback triggering of thyristors/transistors, the stator and rotor field always remain in synchronism as the frequency of triggering automatically adjusts to rotor speed. The length of on-time of thyristors determine the motor torque magnitude. Thus by means of electronic circuitry brushless motors can be controlled for both constant and variable torque operation.

The brushless DC motors, while being generally more expensive for the same kW rating, than commutator and brush motors posses certain advantages over the conventional motors. They are:

(*a*) They require less or no maintenance.

(*b*) They have a much longer operating life.

(*c*) There is no risk of explosion or possibility of RF radiation due to arcing.

(*d*) They produce no brush or commutator particles or gases as byproducts of operation.

(*e*) They are capable of operation submersed in fluids, combustible gases and may even be hermetically sealed.

(*f*) They are generally more efficient than brush-type dc servomotors or conventional dc motors.

(*g*) They provide a more rapid response and a fairly linear output torque vs input current characteristics, which lends itself to servo applications.

4.2.1 Schematic and Operation

The schematic diagram of a brushless dc motor shows the three phases of the stator and rotor with '*d*' and '*q*'-axis. The stator is connected to a variable voltage current source through an inductor and an invertor comprising six SCR's (S_1 to S_6). Diodes are connected across SCR's to protect these from the $L(di/dt)$ voltage induced in the armature coil undergoing commutation. Position sensors placed on the rotor provide signal to the sensor decoders and gate drivers which cause the SCR's to be fired in sequence so as to be in synchronism, with the rotor's mechanical position. The stator and rotor fields thus get locked into each other and remain in synchronism at any rotor speed.

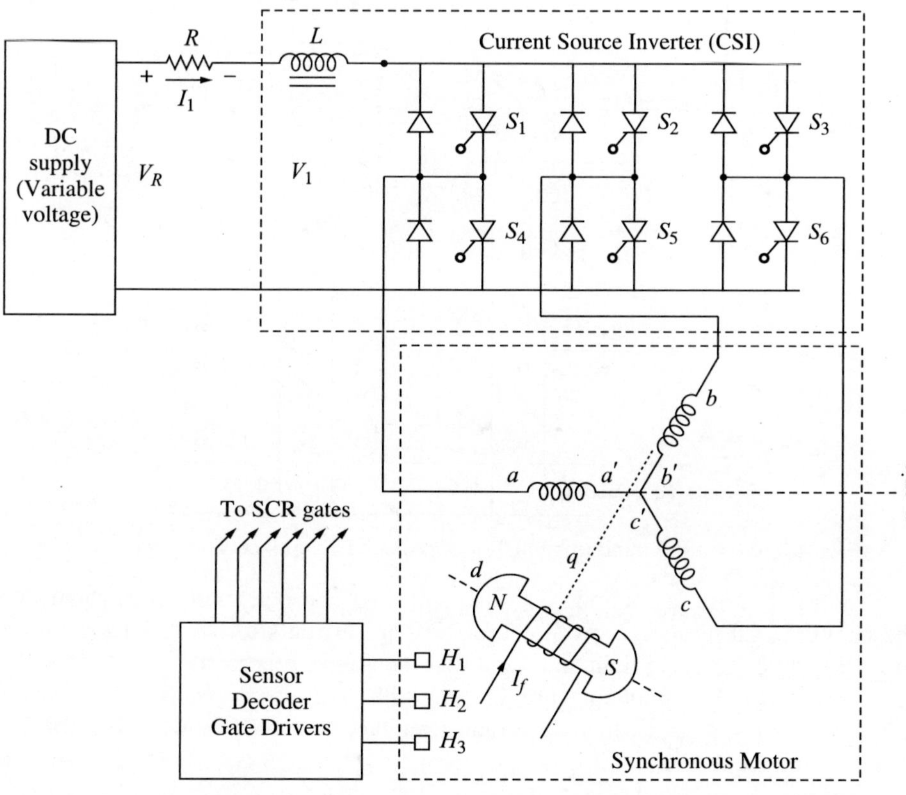

Fig. 4.1 Brussless dc motor: schematic and operation

Ideal phase current are pulses of ±I lasting 120° electric each half and displaced from each other 120° electrical phase to phase. Actual current wave forms differ from the ideal rectangular current waves by gradual rises and falls. Such as inverter where ac current flows in form of constant current pulses is known as Current Servo Inverter (CSI).

Sequence of inverter firing follows from the phase current waveforms. For this sequence of SCR firing 120° or 60° elect spaced sensor codes are generated by means of light sensitive or Hall effect sensors.

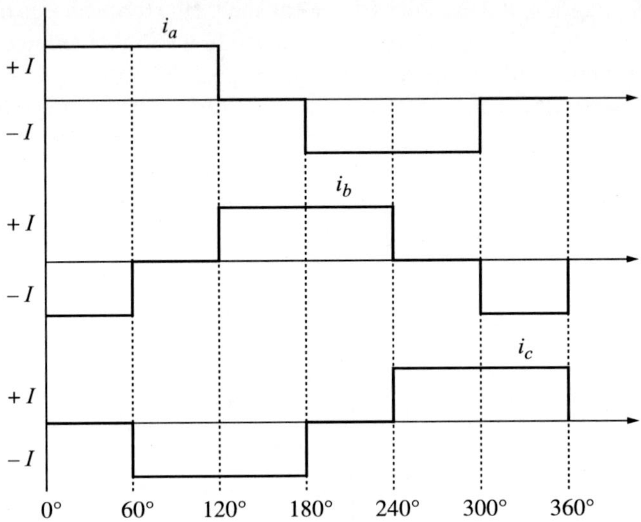

Sensors codes	1	1	1	0	0	0	H_1
120° elect.	0	0	1	1	1	0	H_2
Spacing	1	0	0	0	1	1	H_3
60° elect.	0	0	0	1	1	1	H_1
Spacing	0	0	1	1	1	0	H_2
	0	1	1	1	0	0	H_3
SCR's conducting	1, 5	1, 6	2, 6	2, 4	3, 4	3, 5	

Fig. 4.2 Current waveforms, SCR's conducting sequence and sensor codes

The rotor carries a commutating disc with 180° cut-out, so that as it rotates with the rotor the light sensors receive light for 180° and are dark for 180°. Sensors produce logic '1' while receiving light and logic '0' when dark. It is easily seen that the three sensors (fixed) and commutating disc (rotating with rotor) produce sensor code sequence as given in Fig. 4.3 from which electronic circuitary generates gating pulses for firing SCR's in the sequence.

The relative position of the commutating disc can be adjusted with respect to the rotor poles. For instantaneous rotor position, it is seen that the sensor code is just going to charge from '1 0 1' to '1 0 0'. The phase '*a*' is middle of its current pulse when the current is commuting from phase '*b*' to '*c*'. At this instant the resultant stator field $\vec{F_1}$ is oriented along the *a*-axis.

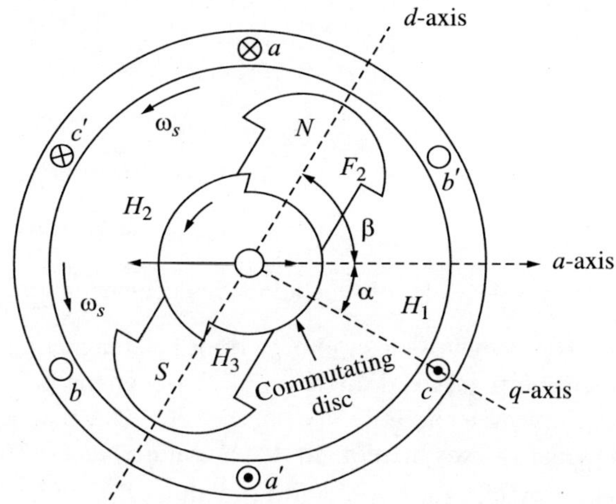

Fig. 4.3

An index marker can be made on the commutating disc which always points to stator north as the stator field rotates in synchronism with the rotor. This index makes an angle 'α' with q-axis of the rotor (lagging d-axis by 90°). Obviously $(\beta + \alpha) = 90°$. The angle may be adjusted by moving the commutating disc on the shaft related to the rotor poles.

4.3 PERMANENT MAGNET DC MOTORS

A number of new permanent magnet materials, ceramics and rare earth magnetic materials have become available commercially. These materials have high residual flux as well as high coercivity. Smaller tractional and sub-tractional hp dc motors are how constructed with permanent magnet poles. As no field windings are needed, so no field current and continuous field loss. As a result PMDC motors are smaller in size than the corresponding rated field wound type motors, this fact partially off-sets the high cost of permanent magnets. These motors offers shunt type characteristic and can only be armature controlled. The risk of permanent magnetism getting destroyed by armature reaction (at starting/reversing or heavy overloads) has been greatly reduced by the new PM materials.

4.4 RELUCTANCE MOTOR

A synchronous reluctance motor is similar to a synchronous AC machine and is described in the section on AC motors. The rotor has salient poles but the stator has smooth, distributed poles where as both the switched and variable machines have salient poles for both the rotor and the stator.

The operating principle of the basic reluctance motor is similar to switched reluctance motor. The synchronous reluctance motor is designed to run on main frequency alternating current and it uses distributed stator winding similar to those used in squirrel cage induction motors. The roles however needs salient poles to create a variable reluctance in the motor's magnetic circuit which depends on the angular position of the rotor.

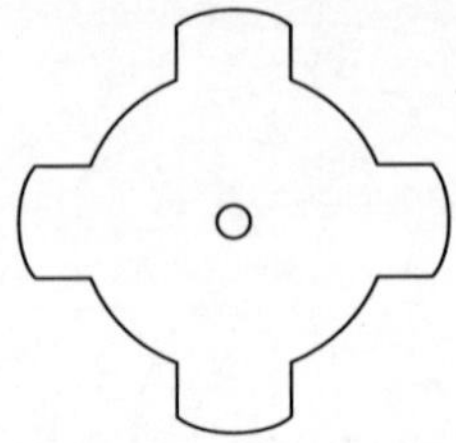

Fig. 4.4 Rotor of synchronous reluctance motor

In a reluctance type synchronous motor, a rotor has magnetic poles which have a low magnetic reluctance in an axial direction of the magnetic poles, and have a high magnetic reluctance toward a circumference of the rotor. The magnetic poles have less components orthogonal to axes thereof, so that armature reaction will be reduced.

The synchronous motor is not self starting without the squirrel cage. During run up it behaves as an induction motor but as it approaches synchronous speed, the reluctance torque takes over and the motor locks into synchronous speed.

4.4.1 Operating Principle

When a piece of magnetic material is free to move in a magnetic field, it will align itself with the magnetic field to minimise the reluctance of the magnetic circuit. To put it another way the piece will orient itself towards the magnetic pole creating the field. The torque on the rotor created in this way is called the reluctance torque.

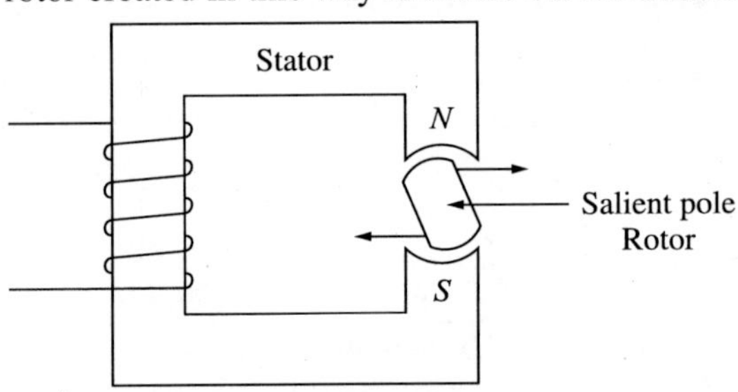

Fig. 4.5 Variable reluctance motor

When the spaces or notches between the rotor poles are opposite to the stator poles the magnetic circuit of the motor has a high magnetic reluctance, but when the rotors poles are aligned with the stator poles the magnetic circuit has a low magnetic reluctance. When a stator pole pair is energized the nearest rotor pole pair will be pulled into alignment with the energized stator poles to minimize the reluctance path through the machine. As with brushless permanent magnet motors, rotary motion is made possible by energizing the stator poles sequentially causing the rotor to stop to the next energized pole.

A polyphase inverter energizes appropriate poles pairs based on shaft position. The excitation of stator poles must be timed precisely to correspond with the rotor position. So that it occurs just as the rotor pole is approaching. The reluctance motor thus requires position feedback to control the motor phase commutation. This feedback control can be provided by using position sensors such as encoders or Hall effect sensors to feedback the rotor angle to trigger the commutator at the appropriate point.

Sensorless position control is also possible at the expense of more complex electronics and software. Motor torque and efficiency are optimised by synchronising the controller switching phase with the rotor position, so that the torque angle is held at its maximum of 90°.

Practical motor designs are doubly salient, with multiple stator and rotor poles. The rotor usually has fewer poles than the stator to enable self-starting and bi-directional control.

Because the rotor is not permanent magnet, it is constructed from iron. Hence no back emf is generated.

Synchronous Reluctance

If the rotating field of large synchronous motor with salient poles is de-energized, it still develop 10 or 15% of synchronous torque. This is due to variable reluctance throughout a rotor revolution. There is no practical application for a large synchronous reluctance motor.

If the slots are cut into the conductor less rotor of an induction motor, corresponding to the stator slots, a synchronous motor results. It starts like an induction motor and runs with a small amount of synchronous torque. The synchronous torque is due to change in reluctance of a magnetic path from the stator through the rotor as the slots align. This motor is an inexpensive. Means of developing a moderate synchronous torque. Low power factor, low pull-out torque and low efficiency are characteristics of the direct power line driven variable reluctance motor.

4.4.2 Torque – δ Characteristics

In synchronous reluctance motor, the speed is constant by varying the load for fixed frequency. From the above Fig. 4.6, by increasing the torque, the motor speed becomes constant. When the torque exceeds the maximum values, the motor goes to out of synchronism.

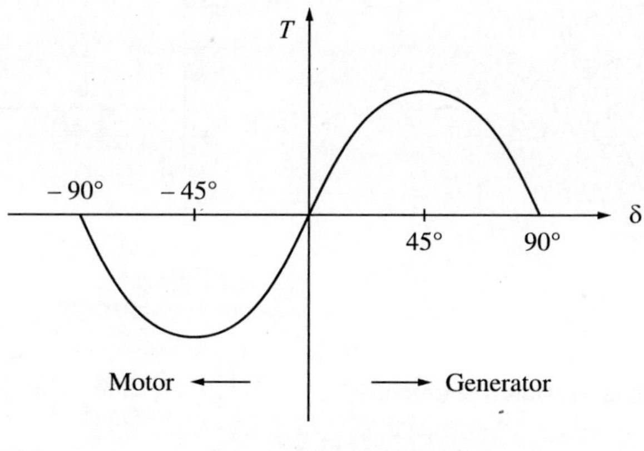

Torque – δ characteristics

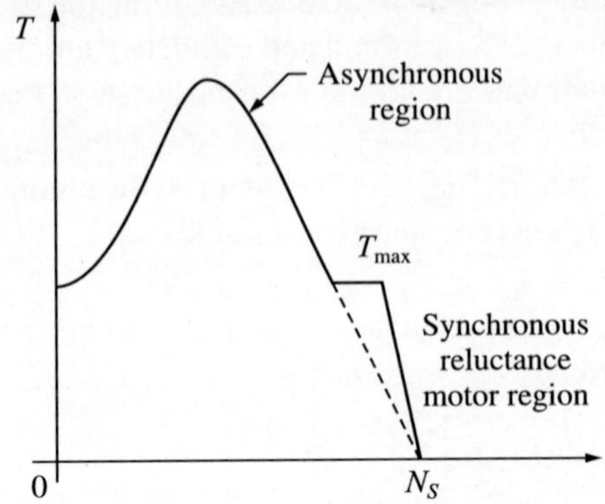

Fig. 4.6 Speed-torque characteristics.

4.5 HYSTERESIS MOTOR

The stator of a hysteresis motor is wound with main and auxiliary windings with a permanently connected capacitor for phase splitting. The capacitor is selected to create balanced 2-phase conditions. The rotor is a smooth solid cylinder of hard steel (this has high hysteresis loss) and does not carry any winding (no rotor bars).

Both the rotor windings are distributed such as to create a rotating field with as nearly a sinusoidally space distribution as possible; this is necessary to keep down iron-loss due to space harmonics of the field.

The phenomenon of hysteresis causes the rotor magnetization to lag behind the stator-created mmf wave. As a consequence, the rotor flux lags by an angle δ the stator mmf axis. Figure 4.7(*a*) shows the magnetic condition in the motor at any instant.

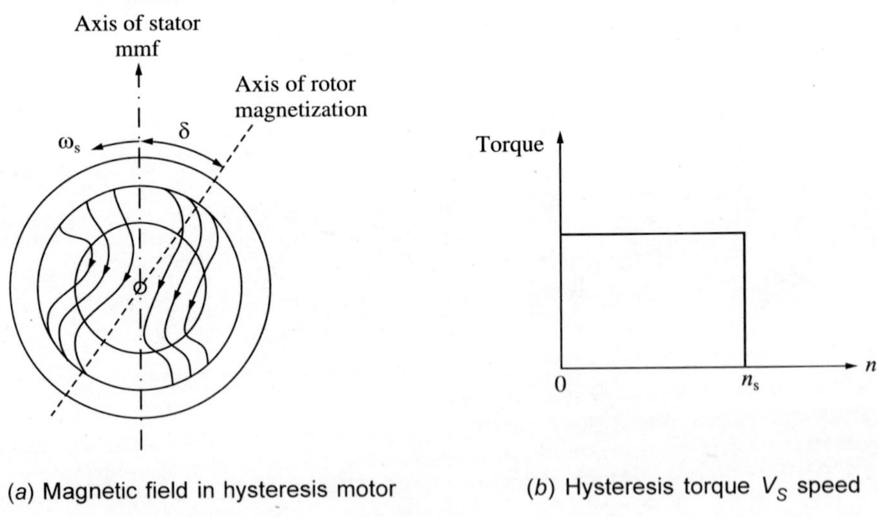

(*a*) Magnetic field in hysteresis motor (*b*) Hysteresis torque V_S speed

Fig. 4.7

As the angle δ is hysteresis-dependant, it remains constant, at all rotor speeds. The interaction torque between stator and rotor field therefore is constant at all speeds. Under the influence of the hysteresis torque the rotor accelerates smoothly and finally runs at synchronous speed with angle δ getting adjusted to the load torque. This is a constant to the "pull-in" phenomenon in a reluctance motor when it synchronizes. Constancy of the hysteresis torque is demonstrated by the derivation given below.

The hysteresis loss is expressed as

$$P_h = k_h f_2 B^x \qquad ...(1)$$

where, B = maximum flux density

$f_2 = Sf$ = rotor frequency.

$$\text{Power across air gap} = \frac{P_h}{S} \qquad ...(2)$$

$$\text{Torque developed} = \frac{P_h}{S\omega_s} = \frac{k_h S_f B^2}{S\omega_s}$$

$$= \frac{k_h f B^2}{\omega_s} = \text{constant.} \qquad ...(3)$$

Another component of torque caused by eddy current loss is simultaneously created in the motor. This can be derived as

$$P_e = k_e f_2^2 B^2 \qquad ...(4)$$

$$= k_e S^2 f^2 B^2$$

$$T = \frac{P_e}{\omega_s} = \left[\frac{k_e f^2 B^2}{\omega_s}\right] S \qquad ...(5)$$

As per Eqn. (5), the eddy current torque is highest at start and reduces linearly with slip vanishing at synchronous speed. This torque component aids the hysteresis torque at starting, endowing excellent starting characteristics to the hysteresis motor. The hysteresis motor has a low noise figure compared to the single phase induction motor such that the load runs at uniform speed. This is because it operates at one speed and nearly balanced 2-phase conditions are not distributed. Further, a smooth rotor greatly aids in low noise performance of this motor. Multispeed operation is easily possible by arranging pole changing of stator windings, the rotor being unwound reacts to create the same number of poles as the stator. As the motor has excellent starting characteristics (starting torque equal to running torque). Therefore, it is well suited to accelerate high-inertia loads.

4.6 PERMANENT MAGNET SYNCHRONOUS MOTORS

Permanent magnet synchronous motors are commonly known as permanent magnet AC motors (PMAC). They are classified according to the nature of voltage induced in the stator as sinusoidally excited and trapezoidally excited. These PMAC motors are commonly known as sinusoidal PMAC and trapezoidal PMAC motors.

A sinusoidal PMAC motor has distributed winding (similar to wound field synchronous motor) in the stator side. It employs rotor geometries such as inset or interior as shown in Fig. 4.8(*a*).

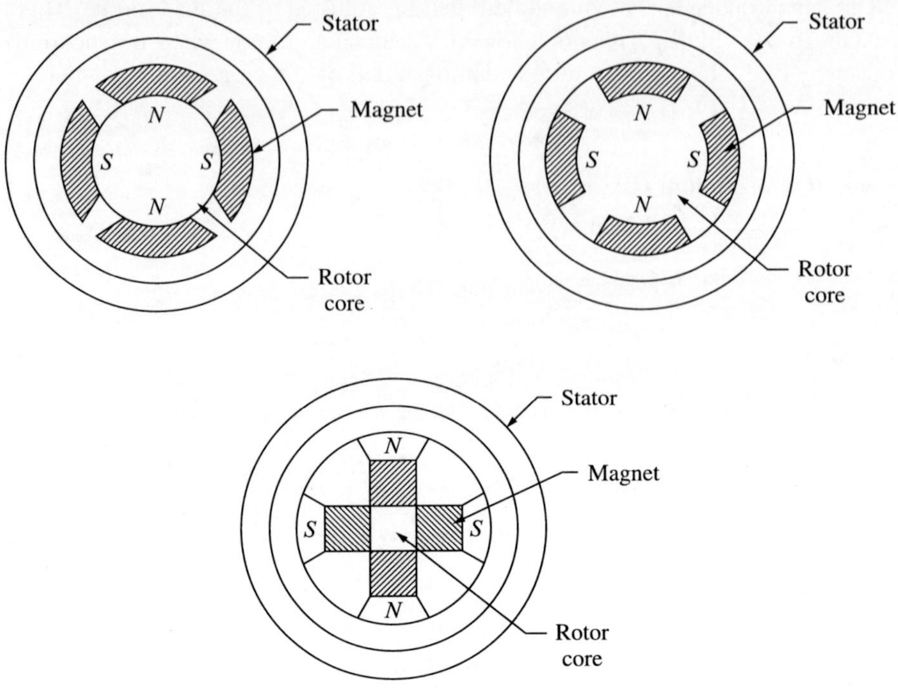

Fig. 4.8 Different types of PMSM

The rotor poles are so shaped that the voltage induced in a stator phase winding has a sinusoidal waveform. The stator of trapezoidal PMAC motor has concentrated windings and a rotor with a wide pole arc.

4.6.1 Sinusoidal PMAC Motors

Since the voltages produced in the stator of a sinusoidal PMAC motors are sinusoidal, ideally the three stator phases must be supplied with variable frequency sinusoidal voltages or currents with a phase difference of 120° between them.

Figure 4.9(a) is the Norton's equivalent of the PMSM equivalent circuit of Fig. 4.8.

where $$\bar{I}_f = \frac{\bar{E}}{jX_s} = \frac{E}{X_s} \angle -(\delta + \pi/2) \qquad \text{...(1)}$$

$$\bar{I}_m = \bar{I}_s + \bar{I}_f \qquad \text{...(2)}$$

The mechanical power developed in the motor is

$$P_m = 3EI_s \cos\left(\delta' - \frac{\pi}{2}\right)$$

Sub for E from Eqn. (1) gives

$$P_m = 3X_s\, I_s\, I_f \sin \delta' \qquad \text{...(3)}$$

$$T = \frac{P_m}{\omega_s} = kI_s\, I_f \sin \delta' \qquad \text{...(4)}$$

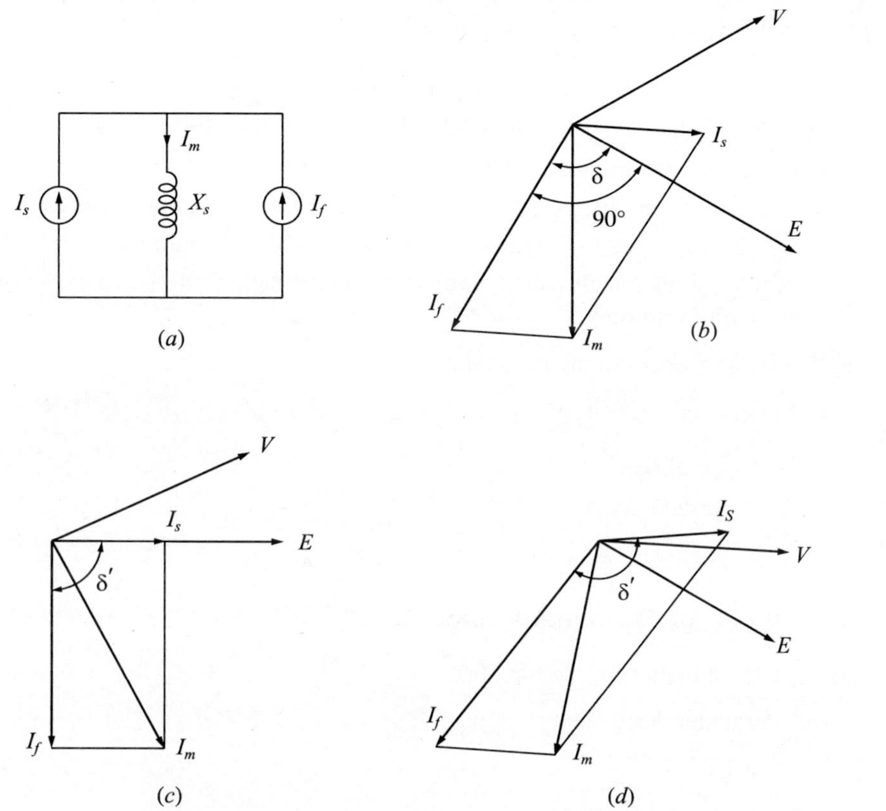

Fig. 4.9 Equivalent circuit and phasor diagrams

Now

where $k = \dfrac{3X_s}{\omega_s}$ = constant

For $\delta' = \pm\, 90°$

$$T = \pm\, kI_f\, I_s = \pm\, k_T\, I_s \qquad ...(5)$$

Hence torque is directly proportional to I_s.

For a given value of I_s, maximum torque can be obtained when $\delta' = \pi/2$. Phasor diagram for $\delta' = \pi/2$ is shown in Fig. 4.9 (*c*).

In this condition motor is said to operate with unity internal power factor because I_s is in phase with excitation emf *E*. The motor itself has a lagging power factor. It is desirable to obtain maximum motor torque per unit of stator current, therefore, this is the preferred operating condition.

These are applications which require speed control in wide range. In wound field motors, the operation up to the base speed is obtained by varying both voltage and frequency. The speed control above the base speed is obtained by reducing the air-gap flux. So that the motor terminal voltage remains at the rated value as frequency is increased.

4.7 PERMANENT MAGNETS

Permanent magnets are magnetic materials that retain their magnetism after being magnetized (retain alignment after removing the external field).

- Magnetostatic energy has been stored in permanent magnets.
- The maximum energy product $(BH)_{max}$ represents the strength of the magnet.
- The strongest hard materials are $Nd - Fe - B$ based materials.
- The transition metals usually provide the high magnetization and the rare-earths the high anisotropy.

Applications of permanent magnets:

- Motors and generators
- Hard disk drive
- Acoustic devices
- MRI

4.7.1 Types of Permanent Magnets

These are three main types of magnets:

(*a*) Permanent magnets
(*b*) Temporary magnets
(*c*) Electromagnets

4.7.2 Permanent Magnets: (Materials Used for Permanent Magnets)

These are four clauses of permanent magnets.

(*i*) Neodymium Iron Boron (NdFeB or NIB)
(*ii*) Samarium Cobalt (SmCo)
(*iii*) Alnico
(*iv*) Ceramic and Ferrite.

This table gives us some of the special characteristics of the four classes of magnets.

Br is the measure of its residual magnetic flux density in Gauss, which is the maximum flux the magnet is able to produce. (1 Gauss is like 6.45 lines/sq. in.).

He is the measure of the coercive magnetic field strength in oersted, or the point of which magnet becomes demagnetized by an external field.

(1 Oersted is like 2.02 ampere-turns/inch).

$\mathbf{BH_{max}}$ is a term of overall energy density. The higher the number, the more powerful the magnet.

$\mathbf{T_{coef}}$ **of Br** is the temperature co-efficient of Br in terms of % per degree centigrade.

This tells you how the magnetic flux changes with respect to temperature. – 0.20 means that if the temperature increases by 100 degrees centigrade, its magnetic flux will decrease by 20%.

Material	*Br*	H_c	BH_{max}	$T_{coeff\ of\ Br}$	T_{max}	T_{curie}
NdFeB	12,800	12,300	40	– 0.12	150	310
SmCo	10,500	9,200	26	– 0.04	300	750
Alnico	12,500	640	5.5	– 0.02	540	860
Ceramic or Ferrite	3,900	3,200	3.5	– 0.20	300	460

Both the Neodymium Iron Boron and the Samarium Cobalt magnets are generally known as rate earth magnets since their compounds come from the rare earth or Lanthanside series 8 the periodic table, of the elements. These are the strongest of the permanent magnets, and are difficult to demagnetize. The T_{max} for the NdFeB is the lowest.

Alnico is made of a compound of aluminium, Nickel and cobalt. Alnico magnets were first developed in 1940's. This magnet is least effected by temperature, but is easily demagnetized. This is the reason why bar paramagnets and horseshoe magnets made of alnico will easily become demagnetized by other magnets, by droping it, and by not storing it with a keeper. Its T_{max}, though is the highest.

Ceramic or Ferrite: Magnets are the most popular types of magnets available today. The flexible magnet we use are the type of ceramic magnet, with the magnetic powders fixed in a flexible binder. These were first developed in 1960's. This is a fairly strong magnet, not as easy to demagnetize as alnico, but its magnetic strength will vary the most as its temperature changes.

The permanent magnets can be made in most any shape imaginable. They can be made into round bars, rectangular bars, horseshoes, rings or donells, disks, rectangular, multi-fingered rings and other custom shapes. Some are cast into a mold and require grinding to achieve final dimensions. Others start as a powder which is pressed into a mold or pressure bonded or sintered.

Basic steps for making a Magnet

1. Make an alloy of iron, boron and neodymium.
2. Grind the alloy into powder.
3. Compress the powder into a shape.
4. Coat the magnet.
5. Magnetize the magnet.

4.7.3 Magnetization Methods

There are two methods generally used to magnetize permanent magnets: static magnetization and pulse magnetization. The first magnetizes via a static electromagnetic field and normally generates a magnetic field of up to 2 MA/m. Pulse magnetization, on the other hand, is used when stronger magnetic fields are desired, or for multipolar magnetization. Each method is optimized for specific materials, shapes and polar configurations.

4.7.4 Magnetization Characteristics

In order to attain maximum magnetic potential, it is necessary to magnetize a material to its saturation point.

- **R Series Samarium Rare Earth Magnets:** Demagnetization characteristics differ depending on the coercivity HcJ of the magnet. Therefore, a greater magnetic field is needed to achieve saturated magnetization for magnets with greater coercivity.
- **N Series Neodymium Rare Earth Magnets:** These magnets are excellent in magnetization, and virtually stable in regard to coercivity. However, special care is needed when re-magnetizing these magnets after demagnetization.

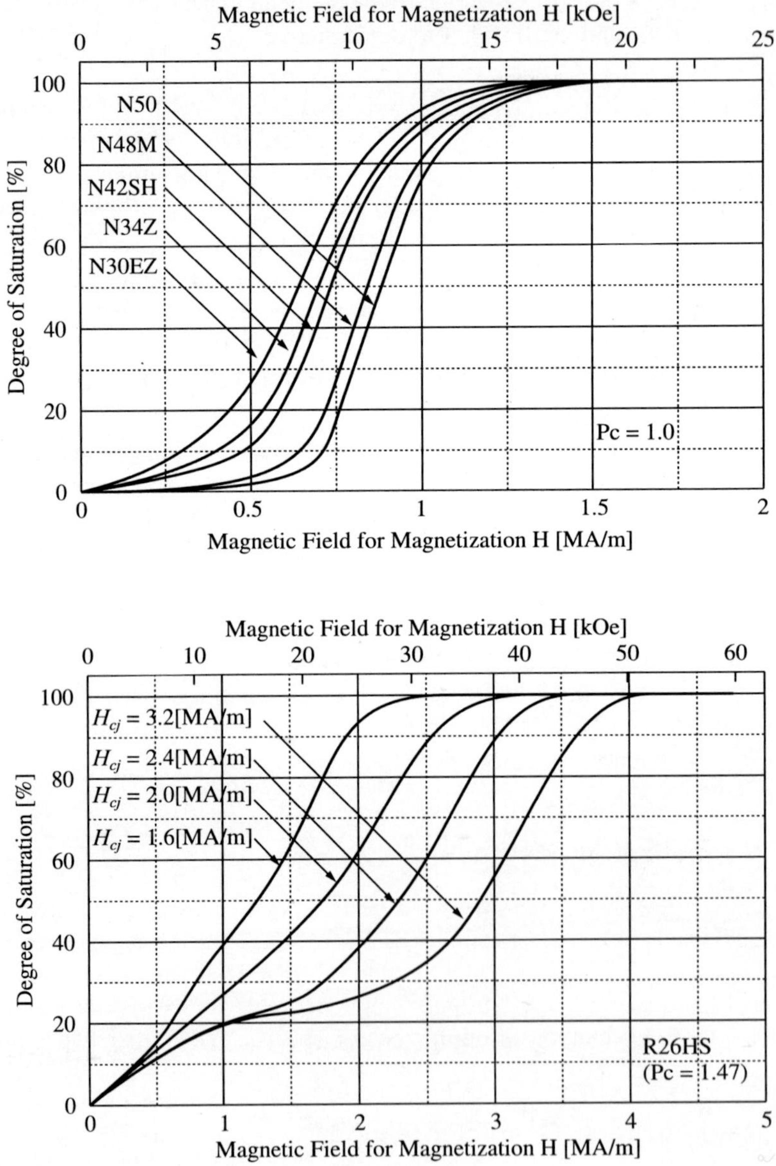

Fig. 4.10

4.8 PCB MOTOR

Operating Principle

PCB motors consist of two parts: The stator, which holds the actuators and electrical connecting circuit; and the rotor, which is pressed onto the surface of the stator and delivers the mechanical output.

A *travelling wave* is generated over the stator surface, acting like a flexible ring to produce elliptical motion on the rotor interface. The elliptical motion of the contact surface propels the rotor and the connected drive-shaft. The teeth attached to the stator can be used to increase rotational speed. Operation depends on friction between the moving rotor and stator as well as amplitude and quality of the wave travelling on the stator.

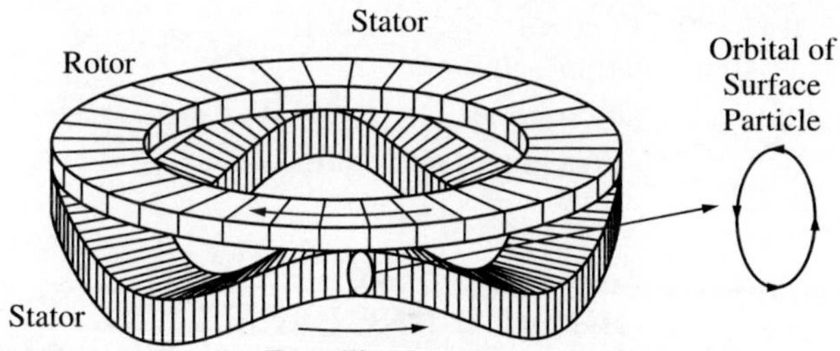

Fig. 4.11

A travelling wave is generated within the stator by activating two modes simultaneously. These modes are induced by a drive circuit performing cos (ωt) and sin (ωt) signals, respectively, close to the mechanical resonance of the stator ring. The travelling wave's direction can be reverse by changing the sign on one of the drive signals.

Rotor

The rectification of the micro-motion is achieved by pressing the rotor on top of the stator. The frictional force between the two causes the rotor to spin. The resulting motion transfer operates as a gear and leads to a much lower rotation speed than the wave frequency. The rotor could be a special ring-shaped material attached to an output shaft, or it could be an integral part of the application that needs to be turned. Since the rotor is a passive component there is a great deal of freedom in the design. Most important is the surface behaviour which must be compatible with the stator teeth material in order to produce the friction needed to drive the application.

Features of PCB Motors

- Direct drive, no gears, slack-less
- High torque and holding torque
- Built-in clutch, tamper proof/safe
- Non-magnetic
- Fast start & stop actions
- Optional position sensor

Applications

- Medico
  - Microscope focus
  - Dosing equipment
- Camera – Pan, Tilt & Zoom
- Instrumentation
  - Dashboard pointers
  - Positioning, stages
  - Laboratory equipment
- Telecom
  - Filter tuning
- Video, and Audio
  - Audio mixing equipment
- Factory automation and robotics
  - Industrial machinery and equipment
- Defense

SOLVED PROBLEMS

Example 4.1

Permanent magnet bruchless dc motor has a torque constant of 0.12 N-m per Amp. Estimate its no load speed in rpm when connected to a 48 V dc supply.

Solution: No load speed in rpm

$$k_f = 0.12 \text{ Nm/A}$$

$$V = k_c\ \omega_{mo} \quad \rightarrow \quad V\text{-Supply voltage}$$

$$\omega_{mo} = \frac{V}{k_c} = \frac{48}{0.12} = 400 \text{ r/s}$$

$$N_0 = 3819 \text{ rpm.}$$

Example 4.2

A permanent magnet dc commutator meter has a stalling form of 1 N-m the stall amount of 5 A. Estimate its no load speed an rpm when this motor is fed from 28 V dc supply.

Solution:

$$T^{st} = 1 \text{ N-m} \qquad I_{st} = 5 \text{ A} \quad \text{and} \quad V = 28 \text{ V}$$

$$V = k_c\ \omega_{mo} \qquad I_{st} = k_f\ I_{st}$$

$$k_f = \frac{T_{st}}{I_{st}} = \frac{1}{5} = 0.2 \text{ Nm/A}$$

$$\omega_m = \frac{V}{k_c} = \frac{28}{0.2} = 140 \text{ r/s}$$

$$= 140 \times \frac{60}{2\pi} = 1336.9 \text{ rpm}$$

Example 4.3

During an overload the motor temperature becomes incisive and the magnet are 20% of the remanent flux density of the armature resistance in 0.8 Ω. Determine the speed of which motor will, when load torque in 0.3 N-m. Assume total brush voltage drop of +V and 28 V a supply ignore friction and windage 10 μm.

Solution: Given data

$$R_a = 0.8\ \Omega \qquad V_{dd} = 2\ \text{V}$$

$$V = 28\ \text{V}$$

$$\phi_2 = 0.8\ \delta_1$$

$$T_{st} = \phi_1\ I_{st} \qquad \text{...(1)}$$

$$T_1 = \phi_2\ I_2 \qquad \text{...(2)}$$

Equation (2) and (1)

$$\frac{T_1}{T_{st}} = \frac{0.8\,\phi_1 I_1}{\phi_1 I_{st}}$$

$$\frac{0.3}{1} = \frac{0.8\,I_L}{5}$$

$$I_L = 1.88\ \text{amps.}$$

$$N_o = \frac{E_{bo}}{k_e\phi_1} \qquad (E_{bo} = V)$$

$$N_L = \frac{(V - I_1 R_a) - V_{dd}}{k_e\ 0.8\,\phi_1}$$

$$\frac{N_L}{N_O} = \frac{V - I_L R_a - V_{dd}}{0.8\phi_1 \times k_e} \cdot \frac{k_e\phi_1}{E_{bo}}$$

$$\frac{N_L}{N_O} = \frac{28 - 1.88 \times 0.8 - 2}{0.8 \times 28}$$

$$\frac{N_L}{1337} = \frac{24.5}{22.4}$$

$$N_L = 1462\ \text{rpm}$$

Example 4.4

A PMBL DC motor has torque constant 0.12 Nm/A. Referred to DC supply find no load speed when connected to 48 DC supply fin state current and state torque if armature resistance = 0.15 Ω, phase and drop in controller transistors is 2V.

Given data:

$$k_e = 0.1\ \text{Nm/A}$$

$$V = 48\ \text{V}$$

Solution: (*a*) No load speed

$$\omega_{mo} = \frac{V}{k_e} = \frac{48}{0.12} = 400\ \text{rad/sec.}$$

$$\omega_{mo} = \frac{2\pi N_o}{60}$$

$$N_o = \frac{480 \times 60}{2\pi} = 3819.71 \text{ rpm}$$

$$N_o = 3819.7 \text{ rpm}$$

(*b*) $R_{ph} = 0.15\ \Omega$

$V_{dd} = 2$ V

Starting current of stall current $= I_{st} = \dfrac{V - V_{dd}}{2\,R_{ph}}$

$$\frac{48-2}{2 \times 0.15} = 153.33 \text{ A}$$

Starting torque of shall torque

$$T_{st} = k_e\, I_{st}$$

$$= 0.12 \times 153.33$$

$$T_{st} = 18.4 \text{ N-m}$$

Example 4.5

A permanent magnet dc commutator motor has a no-load speed of 6000 rpm when connected to a 120 V supply. The armature resistance is 2.5 Ω and rotational and gross lanes may be neglected determine the speed when the supply voltage is 60 V and the torque is 0.5 Nm.

Given data: $V_s = 120$ V, $R_a = 2.5\ \Omega$, $T = 0.5$ N-m

Solution:

$$E = k_e \omega_m$$

$$V = k_e\, \omega_{mo}$$

$$T = k_e\, I$$

$$k_e = \frac{V}{\omega_{mo}} = \frac{120}{2\pi N/60} = \frac{120 \times 60}{2\pi \times 6000} = 0.19$$

$$I = \frac{T}{k_e} = \frac{0.5}{0.19} = 2.617 \text{ A}$$

$$E = V - I_a R_a$$

$$= 60 - 0.617 \times 2.5$$

$$= 53.45 \text{ V}$$

$$\omega_m = \frac{E}{k_e} = \frac{53.45}{0.19} = 281.34 \text{ rad/sec}$$

$$\omega_m = \frac{2\pi N}{60}$$

$$N = \frac{281.34 \times 60}{2\pi} = 2686.6 \text{ rpm}$$

Example 4.6

In an actuator used in a disk drive for personal computer. The magnet length is 6 mm and the airgap in 2 mm. If the magnet is ferrite with a remanent flux density of 0.03 T. Calculate the flux density in the airgap. Assuming in finitely permeability. Iron components use the magnetic flux path is complexity by a self iron bridge which is encircled by a coil of 200 turms mounted on the relating armature, if the current in the coil in 10 mA. Calculate the torque the radial distance from the centre line to the airgap in 25 mm and the height of the active magnetic parts in 50 mm top to bottom.

Given data: $I = 2$ *mm,* $l_m = 6$ *mm*

B_r = *Permanent flux density* = *0.35*

$l = 50$ *mm;* $r = 25$ *mm,* $I = 10$ *mA*

to find torque.

Solution: Flux concentration factor $= \dfrac{A_m}{Ag}$

$$A_m = \frac{2}{3}\pi\left[n - g - \frac{l_m}{2}\right] l$$

$$= \frac{2}{3}\pi\left[25\times10^{-3} - 2\times10^{-3} - 3\times10^{-3}\right]50\times10^{-3}$$

$$= 2.0943 \times 10^{-3}\ \text{m}^2$$

$$A_g = \left[\frac{2}{3}\pi\left(r - \frac{g}{2}\right) + 2g\right](l + 2g)$$

$$\left[\frac{2}{3}\pi(25-1)\times10^{-3} + (2\times2\times10^{-3})\right](50 + 10^{-3} + 2\times2\times10^{-3}) = 2.9303 \times 10^{-3}\ \text{m}^2$$

$$C_\phi = \frac{A_m}{A_g} = \frac{2.0993\times10^{-3}}{2.9303\times10^{-3}} = 0.7146$$

$$P_m = P_{mo} + P_{rl}$$

where

$$P_{rl} = 0$$

$$P_m = P_{mo}$$

$$\therefore \quad P_{mo} = \frac{m_o m_{rec.} \cdot A_m}{l_m}$$

For normal *BLDG* motor, $HC = 606$ kA/M

$$B_m = B_r\left[i - \frac{H_m}{H_C}\right]$$

$$H_m = \frac{NI}{l} = \frac{200\times10\times10^{-3}}{50\times10^{-3}} = 40$$

$$\therefore \quad B_m = 0.3\left[1 - \frac{40}{606\times1.3}\right] = 0.29998$$

$$M_{rec} = \frac{B_r - B_m}{M_o H_m} = \frac{0.3 - 0.29998}{4\pi \times 10^{-7} \times 40}$$

$$= 0.395897$$

$$P_{mo} = \frac{4\pi \times 10^{-7} \times 0.395897 \times 2.0943 \times 10^{-3}}{6 \times 10^{-3}}$$

$$= 1.73652 \times 10^{-7} \text{ Wb/AT}$$

$$R_g = \frac{g'}{m_o A j}$$

$$g' = k_e g$$

where $k_c = 1.05$

$$g' = 1.05 \times 2 \times 10^{-3} = 2.1 \times 10^{-3}$$

$$R_g = \frac{2.1 \times 10^{-3}}{4\pi \times 10^{-7} \times 2.9309 \times 10^{-3}} = 570.286$$

$$B_g = \frac{C_\phi}{1 + P_m R_g} B_r$$

$$= \frac{0.71469 \times 0.3}{1 + (1.7305 \times 10^{-7} \times 570.286)}$$

$$= 0.2144 \; T$$

$$\text{Torque/phase} = 2 T_{ph} B_g l_r 1$$

$$= 2 \times 200 \times 0.2144 \times 50 \times 25 \times 10 \times 10^{-3}$$

$$T = 1.072 \times 10^{-3} \text{ N-m}$$

Example 4.7

AS phase 16 pole synchronous motor has a star connected winding with 144 slots and 10 conductors per slots the flux per pole in 0.03 mb. Sinusoidally distributed and the speed in 375 rpm. Find the frequency and the phase and line emf. Assume full pitched cost.

Solution:

$$f = \frac{PN}{120} = \frac{16 \times 375}{120} = 50 \text{ Hz}$$

k_c is not given, it would be taken as unity (full pitched coil)

$$n = \text{slots per pole} = \frac{144}{16} = 9$$

$$m = \text{slots per pole phase} = \frac{h}{3}$$

$$\beta = \text{slot angle} = \frac{180}{n} = \frac{180°}{9} = 20°$$

$$m = \frac{144}{16/3} = 3$$

$$\text{Distribution factor} \quad k_d = \frac{\sin\left(\text{in}\frac{\beta}{2}\right)}{m\sin\left(\frac{\beta}{2}\right)} = \frac{\sin\left(3\times\frac{20}{2}\right)}{3\sin\left(\frac{20}{2}\right)} = 0.9598$$

$$\text{Total armature conductors } z = \frac{\text{Conductros}}{\text{slots}} \times \text{no. of slots}$$

$$= 10 \times 144 = 1440$$

$$\text{No. of armature per phase conductions} = \frac{1440}{3} = \text{Conductors/Phase (zph)}$$

$$\text{No. of turns/ph} = \frac{Z_{ph}}{2} = \frac{480}{2} = 240$$

$$\text{Emf. induced per phase } E_{ph} = 4.44\ k_c k_d f_\phi T_{ph}$$

$$= 4.44 \times 1 \times 0.96 \times 50 \times 0.03 \times 240$$

$$= 1534 \text{ V}$$

$$\text{Line voltage } V_L = \sqrt{3}\,V_{ph}$$

$$= \sqrt{3} \times 1534 = 2658 \text{ V}$$

Note:

$$T_{ph} = \frac{Z_{ph}}{2}$$

$$T_{ph} = \frac{\text{No. of conductors}}{\text{Slots}} \times \frac{\text{No. of slots}}{\text{Phase}}$$

$$T_{ph} = \frac{10 \times 144}{2 \times 3} = \frac{1440}{6} = 480$$

Example 4.8

A three phase four pole brush less pm rotor has 36 stator slots, each phase winding is made up of three coils per pole with 20 torn per coil. The coil span in seven slots. If the fundamental component of magnet flux is 1.8 m w.b. Calculate the open circuit phase emf (E_g) of 3000 rpm.

Given: 3ϕ, 4 pole, 36 stator slots

3 coils per pole, 20 torns per coil

Coil span = 7 slots

$\phi_m = 1.8\ mb$

$N = 3000$ rpm

Open circuit phase emf $E_q = 1.$

Solution:

$$f = \frac{PN}{120} = \frac{4 \times 3000}{120} = 100 \text{ Hz}.$$

$$T_{\text{ph}} = \text{No. of tarns. of phase}$$

$$= \left(\frac{\text{No. of tarns}}{\text{coils}}\right) \cdot (\text{coil/poles/phase}) \times (\text{No. of poles})$$

$$= 20 \times 3 \times 4 = 240$$

$$R_{d1} = \frac{\sin \frac{m\beta}{2}}{m \sin \frac{\beta}{2}}$$

$$m = \text{Slots per pole per phase.}$$

$$m = \frac{36}{4} \times 3 = 3$$

$$\beta = \text{Slot angle} = 180°/n$$

$$n = \text{Slots per pole} = \frac{36}{4} = 9$$

$$\therefore \quad \beta = \frac{180°}{9} = 20°$$

$$\therefore \quad k_{d1} = \frac{\sin\left(\frac{3 \times 20}{2}\right)}{3 \times \sin \frac{20}{2}} = 0.9558$$

$$\text{Full pitch} = 9 \text{ slots/pole}$$

$$\text{Coil span} = 7 \text{ slots}$$

Short pitched by 2 slots ⇒ 2 × 20° = 40°

$$\alpha = \text{angle of short pitch}$$

$$k_{p1} = \cos \frac{\alpha}{2} = \cos \frac{40}{2} = 0.9396$$

$$k_{s1} = 1$$

∴ Winding factor

$$k_{n1} = k_{d1} k_{p1} k_{s1} = 0.9018$$

Open circuit phase emf is

$$E_q = 4.44 \, f \, \phi_m \, k_{w1} \, T_{\text{ph}} \text{ volts}$$

$$= 4.44 \times 100 \times 1.8 \times 10^{-3} \times 0.9018 \times 240$$

$$= 172.97 \text{ V}$$

$$E_q = 173 \text{ V}$$

REVIEW QUESTIONS

1. What are the advantages and disadvantages of Brughless dc motor drives.
2. List out the various permanent magnet materials.
3. Draw the magnetic equivalent circuit of 2 pole permanent magnet brughless dc motor.
4. Why is a PMBLDC motor called an electronically commutated motors?
5. What are the relative merits of the brughless dc motor.
6. Derive the expressions for the emf and torque of a PM Brughless dc motor.
7. What are the difference between connectional dc motor and PMBLOC motor.
8. What arc the applications of PMBLOC motor?
9. Mention few advantages of permanent magnet brughless DC motor.
10. Sketch torque-speed characteristics of ideal brughless DC motor.
11. Compare PMBLOC motor and switched reluctance motor.
12. Name the position sensors that are used for PMBLOC motor.
13. What is a permanent magnet synchronous motor.
14. What are the difference between connectional synchronous motor and pm synchronous motors?
15. What are the advantages and disadvantages of PMSM.
16. Explain the construction and operation of PMSM.
17. What are the application of PMSM motors.
18. Clearly explain the differences between synchronous reluctance motor and PMSM.
19. What are the difference in the constructional features of PMBLDC and PMSM.
20. Write the emf equation of PMSM.
21. Derive the emf equation of PMSM.
22. Derive the expressions for power input and torque of a PMSM. Explain how its torque speed characteristics is obtained.

❑❑❑

Chapter 5
Single Phase Commutator Motors

5.1 SINGLE PHASE COMMUTATOR MOTORS

A part from series traction motors still in use on low-frequency railways. In Europe, 1-Ph commutator motors are uncommon in rating over 5 kW. Fractional kilowatt motors, however, are made in large numbers.

5.2 BASIC CONNECTION ARE TWO TYPES

1. **Universal motor:** This resembles a dc series motor except that the whole magnetic circuit must be laminated to limit core loss. Small motors for drills and domestic machines can be run on both dc and 1-Ph supplies.
2. **Repulsion motor:** The machine has the rotor energized inductively the rotor winding is designed for a low working voltage and its brushes are joined by a short-circuit connection to provide a direct current path the stator winding has either *d* axis and *q*-axis section.

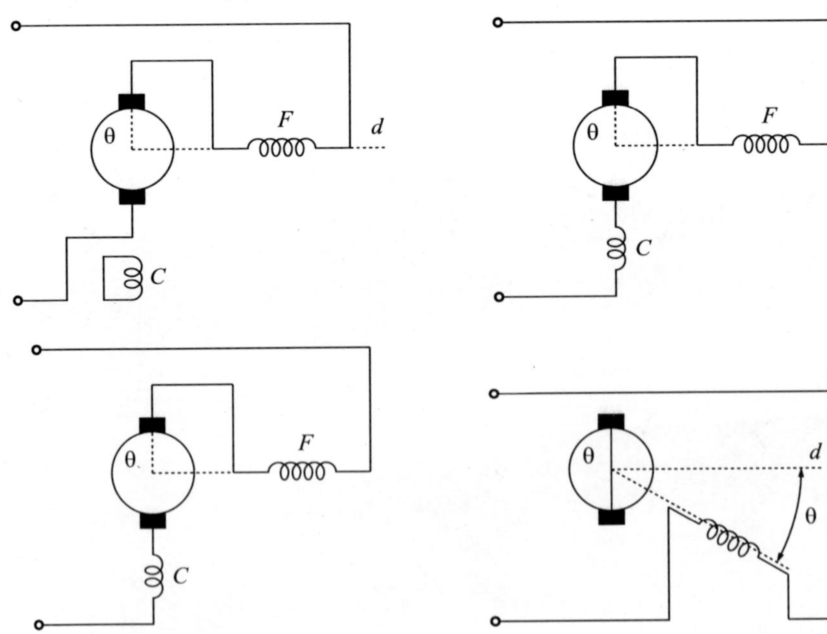

Universal Motor

Repulsion Motor

Fig. 5.1

Universal Motor: The machine has a series torque/speed charge if speed is readily controllable. Typically 3000 – 10000 r/min, a rating up to about 500 W, and a shaft mounted fan for cooling. To reduce winding reactance a 2 or 4 pole structure is employed without compole or neutralizing windings to ameliorate commutation. For in cheap machines (and cast is the competitive factor) with few rotor slots and sectors the reactance and pulsational. Voltage may sum to 12-14 V Brush pressure is high to restrict bounce and brush positioning is crucial to satisfactory commutation.

The rotor winding in slots skewed by one pitch, is conventional. The stator laminations have salient poles shaped to accept machine-wound field coils. Large core sections and short air-gaps are employed to reduce saturation and exciting m.m.f. Production tolerance on rotor eccentricity must be such as to limit unbalanced magnetic pull, which may cause bearing 'hammer'.

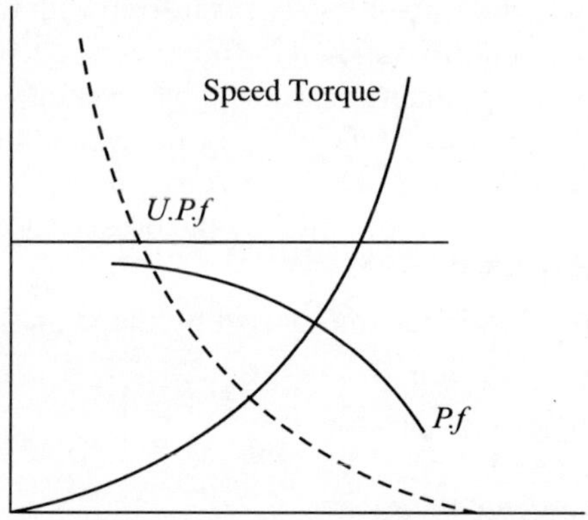

Fig. 5.2 Torque speed characteristic

A series motor can be run from either dc or ac (single phase) supply provided that both stator and rotor cores are laminated to limit the iron-loss. The cross-sectional view of a series motor connected to ac supply. As the armature current alternates, the field polarity alternates in phase with it as shown in Fig. 5.3.

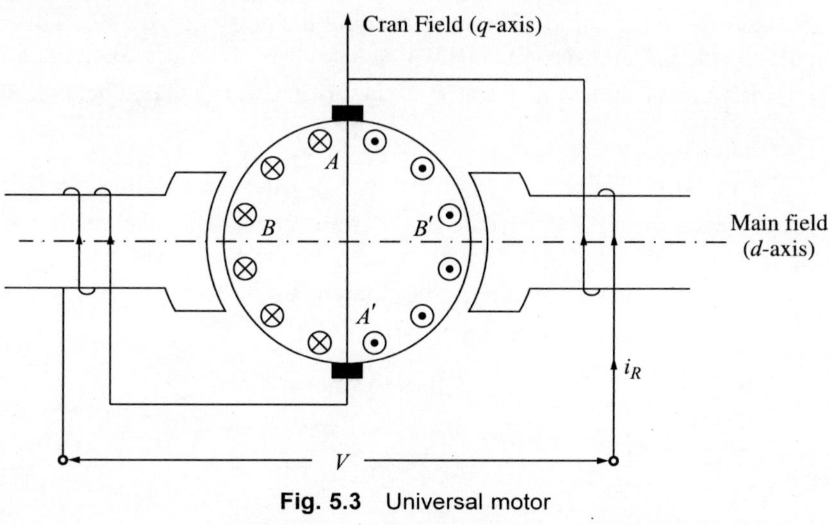

Fig. 5.3 Universal motor

As a consequence the torque developed ($\alpha\ \phi\ ia$) is unidirectional with a pulsating component over and above the average value. The pulsating component of torque is filtered out by the rotor and load inertia so that the speed pulsation are almost negligible. The necessity of laminating the stator (poles and yoke) is indicated by the alternating flux it is required to carry.

The current carrying armature produces cross-flux along the *q*-axis in space quadrature with the main flux (in the *d*-axis) these two fluxes apart from being. In space quadrature are alternating at a supply frequency each of these fluxes then induces rotational and static emf in the armature.

Emf's of Main Field (*d*-axis)

As in the dc case, rotational emf is induced in armature conductor adding to a maximum value of the *q*-axis brushes, the emf alternates in uniform with flux/pole. The magnitude of this emf is maximum in coil BB' shown in figure and is zero in coils AA' under the brushes. Hence commutation remains unaffected by this emf caused by time-variation of the flux/pole transformer. Emf's are induced in armature coils which add to net. Zero value of the brushes. There emf's have zero value in coils BB' and maximum value in coil AA'. Since coils AA' are started by the brushes. Commutation is seriously affected resulting in brush sparking.

The rms value of the rotational emf caused by the main field is

$$E_a = \frac{1}{\sqrt{2}} \frac{\phi n Z}{60} \times \left(\frac{P}{A}\right)$$

where ϕ = maximum value of flux/pole.

This emf is in-phase with the main flux phasor.

Emf's of Cross field (*q*-axis field)

The rotational emf caused by cross flux (along *q*-axis) induces net zero emf at the brushes. This emf is zero in coil BB' and maximum in coils AA' thereby interfacing in the commutation process. On the other hand, the transformer emf of armature coils adds to a definite value and is alternating leading the cross flux by 90°. This emf is maximum in coil BB' and zero in coils AA' and so does not affect commutation.

The rms value of transformer emf E_t at the brushes is calculated below.

$$E_t = \sqrt{2}\ Zk_b f \phi_a N_p$$

where N_p = turns/parallel path, ϕ_a = armature (iron) flux, proportional to I_a

k_b = space factor: though coil emf's are in phase their magnitude varies as cosine of angle the coil axis makes with the brush axis.

Assuming the armature winding to be finely distributed, it can be easily shown that

$$k_b = \frac{2}{\pi}$$

hence

$$E_t = 2\sqrt{2} f \phi_a N_p$$

The ratio of the transformer emf (E_t) to the rotational emf (E_a) is

$$\frac{E_t}{E_a} = \frac{2\sqrt{2} f \phi_a N_p}{\frac{1}{\sqrt{2}} \frac{\phi_n Z}{60} \left(\frac{P}{A}\right)}$$

But $$N_p = \frac{Z}{2A}$$

$$\frac{E_t}{E_a} = \frac{120 f \phi_a}{n P \phi} = \frac{ns}{n} \cdot \frac{\phi_a}{\phi}$$

where $$n_s = \frac{120 f}{P} = \text{synchronous speed.}$$

Torque Developed

The torque developed is pulsating in value with an average value and a predominant second harmonic component. The average torque is given by

$$T_{\text{av}} W_m = E_a I_a$$

Being a series motor $E_a \propto I_a$ (assuming linear magnetic circuit). The torque is therefore proportional to square of armature current which becomes directly proportional of high armature current because of saturation of the magnetic circuit.

Performance Characteristics

The performance characteristics of ac operated series motor. The speed-torque characteristic has a typical series shape. The speed-torque characteristics for dc operation would lie. Some what higher as shown dotted, this is because of reactance voltage drops $[I_a (x_f + x_a)]$ in ac operation so that E_a and hence speed is lower for a given current and torque the power factor of an ac operated series motor is rather poor because of the large series reactance $(x_f + x_a)$.

The no-load speed of a universal motor unlike other machines is very high (of the order of 2000 rpm) therefore, the motor is smaller in size than other types for a given load. Universal motors are used where light weight is important as in vacuum cleaners and portable tools which usually operate of high speeds (1500 – 15000 rpm).

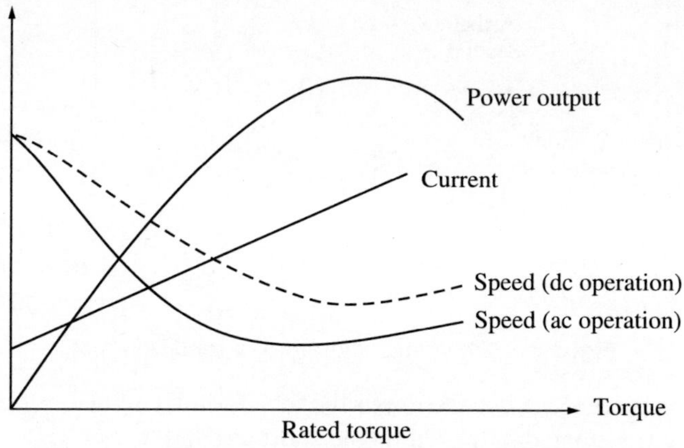

Fig. 5.4 Performance characteristics of universal motor

Compensating and Inter-pole Windings

To improve the power factor of a series motor, the cross-flux which is mainly responsible for armature reactance must be cancelled out by a compensating winding connected in series with the armature circuit. The axis of this winding is along the brush axis and the winding must be spread over the full pole pitch for effective neutralization of cross-flux like in a dc machine the interpole winding when provided must be effective over a narrow interpolar region. The schematic connection diagram of all the stator windings and the armature is shown in Fig. 5.5.

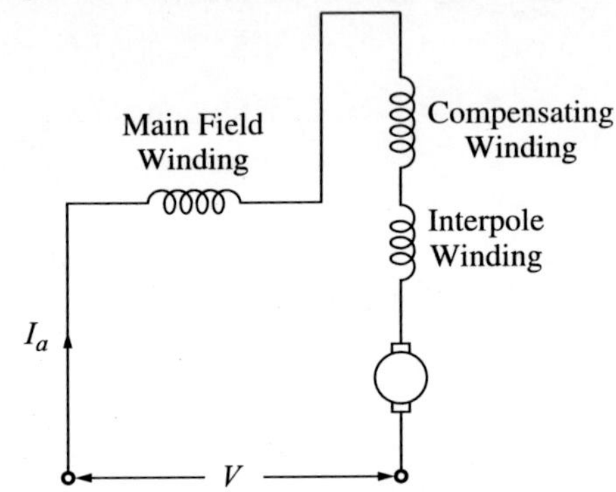

Fig. 5.5 Arrangement of stator winding in ac operated series motor

5.3 REPULSION MOTORS

The repulsion motor is like a dc series motor with the rotor energized inductively the stator sloty house a single-phase exciting winding similar to the main winding of a single-phase induction motor. The rotor carries an ordinary distributed dc winding connected to the commutator of one end. The brushes are short-circuited to provide a closed current path as shown in Fig. 5.6.

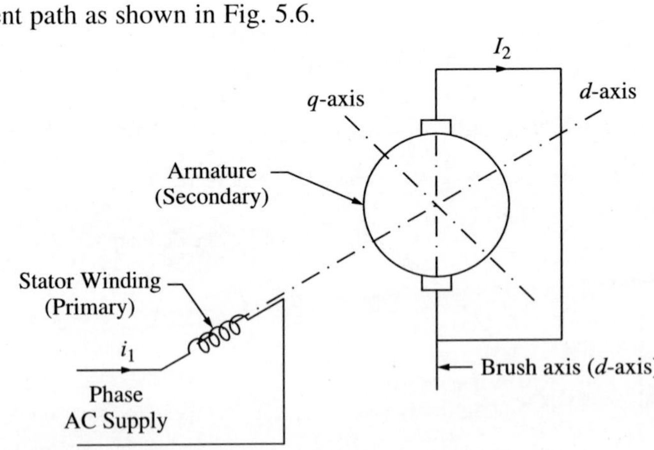

Fig. 5.6 Schematic diagram of a repulsion motor

If the brushes are located in the d-axis the transformer emf in the effective armature coil will be maximum and so would be the short circuit current I_2, but no torque will be produced as stator and rotor fields are aligned. If the brushes are placed in the q-axis. The transformer emf induced in the armature would add up to zero and so current $I_2 = 0$ and hence no torque is produced. As it is essential that both an armature current should flow and an angular displacement must exist between the two fields the brushes are placed in an intermediate position making a large angle of about 70° with the *d*-axis. In a fertile attempt to reduce copper losses the rotor runs in a direction to orient itself with the *q*-axis, *i.e.* the direction of rotation is counter clockwise for brushes displaced from the *d*-axis shown in Fig. 5.6.

This appears title repulsion between the two fields and hence the same. The direction of rotation will reverse if the brushes are displaced on the other side of the *d*-axis.

The unmodified repulsion motor has speed-torque characteristic similar to that of an ac series motor. It has a large starting torque and a tendency to over speed of light load.

5.4 LINEAR INDUCTION MOTORS

5.4.1 Introduction

A linear motion can be mechanically obtained from a rotating electric motor. By linear motion we mean motion in a horizontal direction. A force producing linear motor is called thrust, propulsion force or traction force. However, linear electric motors (LEMs) have been developed which can provide linear motion much more economically and efficiently.

In principle, each rotating electric machine can have its counter part in the form of LEM. However, only linear induction machines and linear synchronous machines have been found economically variable machines.

5.4.2 Construction of Linear Induction Motor

Figure 5.7(*a*) shows a polyphase rotary induction motor. If this induction motor is cut at *X* and laid out flat as shown in Fig. 5.7(*b*), a linear induction motor is obtained.

In Fig. 5.7(*b*) rotor is shown in the form of conducting sheet. Hence forth, stator and rotor would be called primary and secondary respectively. In Fig. 5.7(*b*) the primary consisting of a core and winding, has a finite length called active length of LIM.

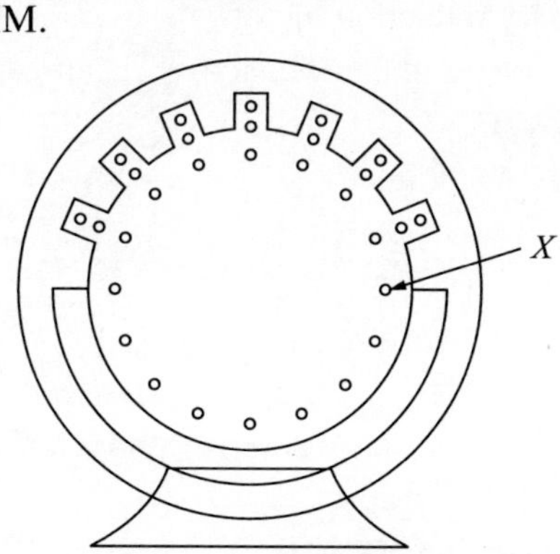

(*a*) Polyphase rotary induction motor

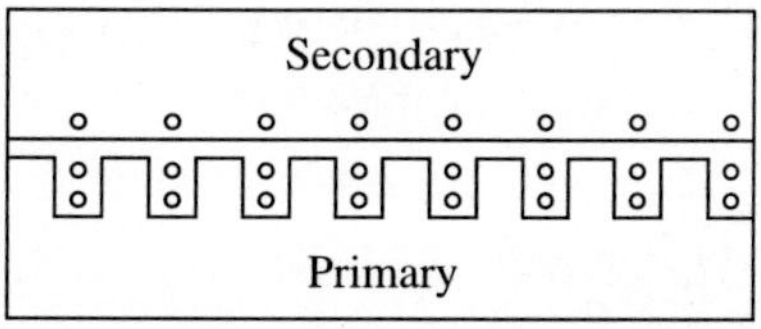

(*b*) Deriving a LIM from rotary induction

Fig. 5.7

5.4.3 Principle of Operation

When the 3-phase winding of Fig. 5.8(*a*) is energized from 3-ϕ supply, a linear travelling field is produced. The velocity of this travelling field is

$$v_s = 2f \text{ (pole-pitch)} = f\lambda \text{ m/sec}$$

where λ is the wavelength of the travelling field and is equal to two poles pitches.

This linear travelling field induces current in the secondary. The interaction between primary and secondary fields results in the production of linear force. In an ordinary induction motor, the speed of the secondary in LIM is less than v_s and is given by equation

$$v_r = v_s(1 - s) \text{ m/sec}$$

Unlike oratory induction motor, a LIM may have moving primary and fixed secondary or moving secondary with fixed primary.

Depending on their use, a LIM may have short primary or a short secondary is shown in Fig. 5.8(*a*) and Fig. 5.8(*b*) respectively.

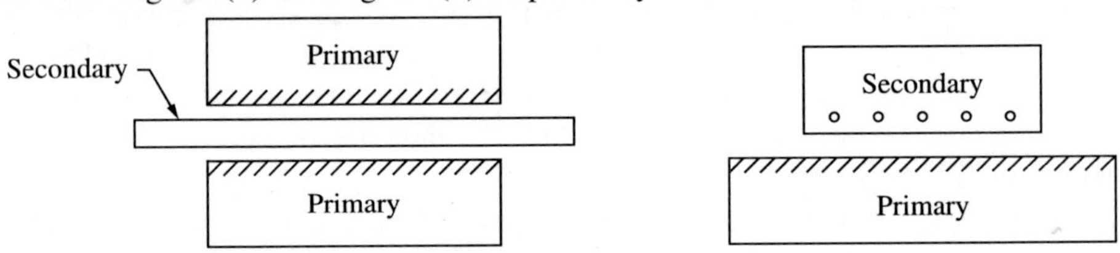

(*a*) Short primary DSLIM (*b*) Short secondary SLIM

Fig. 5.8

LIM may have two primaries to obtain a double-sides LIM (DSLIM) as shown in Fig. 5.8(*a*).

In case LIM has one primary, then it is called single-sided LIM (SLIM).

Linear Force

An approximate expression for the linear force developed in a LIM can be obtained in terms of its main dimensions, just like on ordinary induction motor.

Neglecting magnetizing current and primary losses, the power across the air gap is

$$P_g = 3E_1 I_1 \cos\theta_1 \quad ...(1)$$

where E_1 = per phase emf induced in primary winding.

$$E_1 = \sqrt{2}\pi f k_w N_{ph} \phi$$

I_1 = per phase primary current

$\cos\theta_1$ = primary power factor

The specific electric loading $\overline{ac}$ is defined as the primary ampere conductors per meter length of field system.

$$\overline{ac} = \frac{3 \times N_{ph} I_1 \times 2}{h} \quad ...(2)$$

where $L = PY$ is the total length of the field system along the direction of linear motion shown in Fig. 5.9. Where Y = pole pitch.

The specific magnetic loading B_{av} is defined as the average air flux density in Tesla.

$$\therefore \quad B_{av} = \frac{P\phi}{LW}$$

$$P\phi = V_{av} LW \quad \ldots(3)$$

where W = width of the field system of linear motor as shown in Fig. 5.9.

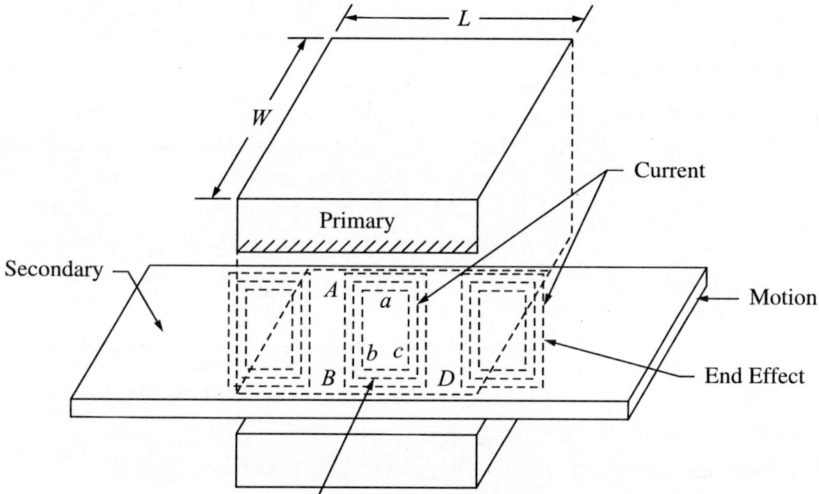

Fig. 5.9 Transverse edge effect and end effect in LEM

Substituting the values of E_1, I_1 in P_g, we get

$$P_g = 3.\sqrt{2}\pi\, k_w\, N_{ph}\, \phi f \frac{\overline{ac} \cdot L}{3 N_{ph} \times 2} \cdot \cos\theta_1$$

We know $\quad f = \dfrac{v_s}{2Y}$

$$\therefore \quad P_g = 3 \times \sqrt{2}\,\pi\, k_w\, \phi \frac{v_s}{2Y} \times \frac{\overline{ac} \times L}{6} \cos\theta_1$$

Now $\quad \phi = \dfrac{B_{av} \cdot LW}{P}$

$$P_g = \frac{\pi}{2\sqrt{2}} \cdot k_w \frac{B_{av} \cdot LW}{P} \cdot \frac{v_s}{Y} \overline{ac} \cdot L \cdot \cos\theta_1$$

Since $L = PY$ and assuming an efficiency of η, the thrust or linear force, developed is

$$F = n \frac{\text{air gap power}}{\text{linear synchronous velocity}}$$

$$F = \eta \frac{P_g}{v_s} = \eta \frac{\pi}{2\sqrt{2}} B_{av} \overline{ac}.LW\, k_w \cos\theta_1 \text{ newtwon}$$

5.4.4 Phenomena Peculiar to LEM

Certain peculiar to LEM must be investigated for better understanding of this motor.

The phenomena which need considerations are:

(*i*) Transverse edge effects

(*ii*) End effects.

These are described briefly.

(i) Transverse edge effects

In an ordinary induction motor, the rotor winding provides well defined path for the current induced in it. In LIM, the secondary consists of a solid conducting plate. Therefore, there are no well defined paths for the currents induced in the secondary. These path is shown in Fig. 5.9. For LIM, the current path perpendicular to the direction of motion such as *ab* and others parallel to *ab*, contribute to the production of useful linear force or thrust. The current paths along the direction of motion, such as *bc*, *BD* etc. contribute towards losses.

The main consequence of transverse edge effects in LEM are as follows:

(*a*) Current paths along the direction of motion add only to the losses of LEM.

(*b*) The effective width *W* of LEM is reduced.

(*c*) Both the above effects deteriorate the LEM performance.

5.4.5 End Effects

Since the primary winding and primary core are of finite length, the end effects appear in LEM.

For example: The field flux the end has different configuration than at the middle. The current induced in the secondary nearer each end go beyond the field structure length *L* as shown in Fig. 5.9. These currents may be termed end-effect currents and they produce additional forces causing braking action specially at lower values of slip.

The consequences of end effects are the production of braking forces in the low slip operation. This has the effect of reducing the efficiency, power factor and output thrust of the LEM.

Applications

From application point of view, LEM's can be divided into three types.

(*i*) force machines

(*ii*) power machines

(*iii*) energy machines

(*i*) Force Machines

These are short-duty machines and operate usually at low speeds. In these machines efficiency is not a major consideration because, these are most useful in small sizes.

Example: Shuttle propulsion, for stop valves, as induction stirrers for molten metals, etc.

(*ii*) Power Machines

These are continuous-duty machines and often operate at medium and high speeds. Power machines must have high operating efficiency.

Example: Conveyors, haulers, electromagnetic pumps, travelling cranes and high speed rail traction.

(*iii*) Energy Machines

These are also short-duty machines. One of the earliest application of LIM as an energy machine was to launch an aircraft. This LIM, called electropull,, was developed in U.S. other modern applications of LIMs as energy machines are as follows:

(*a*) Accelerators for very high velocity projectiles, for exp. for accelerating cars or lorries for indoor crash tests.

(*b*) Actuator for h.v. circuit breakers.

(*c*) Impact extruders for metals.

❑❑❑

Model Paper-1

Time: Three hours *Maximum: 100 marks*

PART-A **(10 × 2 = 20 Marks)**

1. What are the types of synchronous reluctane motor?

2. What are the main features of stepper motor which are responsible for its wide spread use?

3. Define the term stepping angle.

4. Draw the speed-torque characteristics of switch reluctance motor.

5. What is the working principle of switched reluctance motor?

6. Give the merits of PMBLDC motor compare to conventional motor.

7. What is meant by peak recovery current in PMBLDC motor.

8. Synchronous machines with surface-mount magnets have very little difference between direct axis and quadrature axis inductance. Why?

9. What is the magnitude of stator current in PMSM to achieve demagnetization.

PART-B **(5 × 16 = 80 Marks)**

10. (*a*) Describe the constructional details and working principle of synchronous reluctance motor.

OR

(*i*) Draw and explain a typical torque-speed characteristics of a synchronous reluctance motor.

(*ii*) Describe the operation of a variable reluctance type stepper motor.

Also explain about micro stepping.

11. (*i*) Explain the working of hybrid motor.

(*ii*) Explain torque versus stepping rate characteristics of a stepper motor. Also explain about slow range and ramping.

12. (*i*) What is stepping angle? Calculate the stepping angle for a 3-phase 24-pole permanent magnet type stepper motor.

OR

(*ii*) Explain with neat diagram the constructional details and working principle of switched reluctance motors.

13. (*i*) Write a note on the power conductors need in switched reluctance motor?

OR

(*ii*) Explain the principle of operation of PMBLDC motors.

14. (*i*) Drive the emf equation of PMBLDC motor.

OR

(*ii*) Draw and explain the speed-torque characteristics of PMBLDC motor.

15. (*i*) Drive the torque equation in PMSM.

(*ii*) Explain with phasor diagram, the measurement of L_d and L_q in PMSM.

OR

Explain the speed-torque characteristics in PMSM.

Model Paper-2

Time: Three hours *Maximum: 100 marks*

PART-A **(10 × 2 = 20 Marks)**

1. What are the design considerations in synchronous reluctance motor?
2. Draw the steady state phasor diagram of synchronous motor.
3. Sketch the *C*-Dump converter circuit for switched reluctance motor.
4. Write down the expresion for self and synchronous reluctance of PM synchronous motor.
5. What is meant by field oriented canoot of PM synchronous motor?
6. Draw the speed-torque characteristics of ideal brushless dc motor.
7. Draw the magnetic equivalent circuit of 2 pole PMBLDC motor.
8. What are the differences between stepper motors and switched reluctance motors?
9. What is mean by micro stepping in stepper motor?
10. State the operation criteria for which a VRM drive system are designed?

PART-B **(5 × 16 = 80 Marks)**

11. (*a*) Explain the principle of operation of synchronous reluctance motor and derive the torque equation.

OR

(*b*) (*i*) Draw the steady state phasor diagram of synchronous reluctance motor. What changes will be made in the phasor diagram: if a permanent magnet is placed on the rotor of same?

(*ii*) Explain the construction and operation of 3-phase SRM with suitable diagrams.

12. (*a*) (*i*) Derive the torque equation and develop the mechanical characteristics of SRM.

OR

(*b*) (*i*) Show the power output of PMILDC motor is more than the PM synchronous motor for the same size.

(*ii*) A 3-phase 4 pole brushless PM motor has 36 stator slots each phase. Windings is made up of 3 coils/pole with 20 turns/coil. The coil span is 7 slots. If the fundamental component of magnetic flux is 1.8 m. Calculate the open circuit phase EMF of 3000 RPM.

13. (*a*) (*i*) Explain the principle of a sinewave PMSM. Draw the phasor diagram and derive the torque equation.

(*ii*) Describe in detail vector control of PMSM.

14. (*a*) (*i*) Why is a PMBL DC motor called an electronically commutated motor?

(*ii*) Explain the power controller used into PMBLDC motor.

15. (*a*) (*i*) Explain the reason for maximum mean torque in 4 phase motor when the lead angle is 1.5 steps.

(*ii*) Describe the dynamics characteristics of VR stepper motor.

Model Paper-3

Time: Three hours *Maximum: 100 marks*

PART-A **(10 × 2 = 20 Marks)**

1. State the operating criteria for which a VRM drive system are designed.
2. What are the main effects of saturation on VRM performance?
3. State some application of stepping motors.
4. What are different modes of excitation in VRM?
5. Explain cogging in PM DC motor.
6. What are the important problems peculiar to LIM?
7. Draw the phasor diagram of a PM synchronous motor.
8. Define power output ratio of PM SM.
9. State the two classification of PM synchronous machines and the types in each.
10. What is a disc motor and mention its application?

PART-B **(16 × 5 = 80 Marks)**

11. Interior PMSM's are preferred for high $\frac{L_q}{L_d}$ ratio, where would such a feature find application?

OR

12. Commutation-torque ripple pass a serious problem in a PMBDCM
    (*a*) only at low speed
    (*b*) only at high speed
    (*c*) all speed; which is true.
13. What is meant by castlation technique and its application in VRM?

OR

14. State the advantage of hybrid stepping motor.
15. Explain sensorless control in PMSM.

OR

16. What are moving coil motors and their classification?
17. State the essential properties required for a material to construct.

OR

Permanent magnet used in rotating machines. Name two materials.

18. What are the main difference between mechanical and electronic commutation?

OR

19. How the LIM are classified?
20. Explain vector control strategy schemes in PMSM.

Model Paper-4

Time: Three hours *Maximum: 100 marks*

PART-A (10 × 2 = 20 Marks)

1. Mention any two advantages of switched reluctance motors.
2. What is meant by reluctance torque?
3. Draw the speed-torque characteristics switched reluctance motor.
4. Write the application of switched reluctance motor.
5. What is self-control of PMSM?
6. Differentiate electronic commutator and mechanical commutator.
7. What is hall sensor?
8. Define holding torque.
9. What do you understand by full step and half step?
10. Draw the dynamic characteristics of stepper motors.

PART-B (5 × 16 = 80 Marks)

11. Explain the constructional details and principle of operation of synchronous reluctance motors with heat diagrams.
12. (*a*) Explain the speed torque characteristics of synchronous reluctance motor.
(*b*) Draw the phasor diagram of synchronous reluctance motor.
13. Draw the cross-sectional view of switched reluctance motor and explain the principle of operation.

OR

14. Draw and explain the operating modes of any two power converters used for SRM.
15. (*a*) Drive EMF and torque equation of PMSM.
(*b*) Draw and explain the phasor diagram of PMSM.
16. (*a*) Explain vector control of PMSM.
(*b*) Draw and explain the speed-torque characteristics of PMSM.
17. (*a*) Discuss the merits and demerits of PMBLDC motor.

OR

18. Explain the operation of 3-phase bipolar BLDC motor drive with neat diagrams and switching tables.
19. Explain the modes of operation of VR stepper motor.

OR

20. (*a*) Draw and explain the drive circuit for stepper motors.
(*b*) Explain open loop and closed loop control of stepper motor.

Model Paper-5

Time: Three hours *Maximum: 100 marks*

PART-A **(10 × 2 = 20 Marks)**

1. What is vernier motor?
2. Write down any two properties of synchronous reluctance motor.
3. What is holding torque in stepping motors?
4. What is hybrid motor?
5. List out any four application of switched reluctance motors?
6. What are the types of power controllers used for switched reluctance motors.
7. What are the types of permanent magnet DC motors?
8. What is permanent magnet DC commutator motor?
9. List out the differences between the PM brushless DC motors and PM synchronous motors.
10. What is meant by slotless motor?

PART-A **(5 × 16 = 80 Marks)**

11. (*a*) (*i*) Explain the principle of operation and constructional features of synchronous reluctance motor.

(*ii*) Explain the working of vernier motor with a neat diagram.

OR

(*b*) (*i*) Explain circle diagram and torque speed characteristics of synchronous reluctance motors.

(*ii*) Explain the synchronous reactance of PM synchronous reluctance motor in detail.

12. (*a*) Explain the construction and principle of operation of VR stepping motor in detail.

OR

(*b*) (*i*) Write short notes about the drive circuits of stepping motors?

(*ii*) Explain the principle of operation and constructional features of hybrid motor in detail.

13. (*a*) (*i*) Draw and explain the characteristics of SR motors in detail.

(*ii*) Explain the shaft position sensing of SR motor.

OR

(*b*) (*i*) Explain the constructional feature of SR motors in detail.

(*ii*) Explain the role of computers in the control of SR motors.

14. (*a*) Derive the torque and EMF equation of the permanent magnet brushless DC motor.

(*ii*) Explain the principle of operation of permanent magnet brushless DC motor.

OR

(*b*) Explain the modes of operation of power controller for permanent magnet brushless DC motor with a neat diagram.

15. (*a*) (*i*) Explain the principle of operation of permanent magnetic synchronous motors in detail.

(*ii*) Derive the EMF and torque equation of permanent magnet synchronous motors.

OR

(*b*) Explain the microprocessor based control of permanent magnet synchronous motors with a neat block diagram and list out its advantages.

B.TECH.

(SEM. VI) THEORY EXAMINATION 2010-11

SPECIAL ELECTRICAL MACHINES

EEE-012

Time: 3 hours *Total Marks: 100*

Note: Answer all the **Five** questions. All questions carry equal marks.

1. Answer any **four** parts of the following: **(5 × 4 = 20)**

(*a*) Explain, how the equivalent circuit of an ordinary polyphase induction motor is also applicable to deep-bar induction motors.

(*b*) In a double-cage polyphase induction motors, explain how the desirable features of high-starting torque and low-operating slip are attained.

(*c*) Discuss the relative merits and disadvantages of single-cage and double-cage induction motors.

(*d*) Show that the maximum electromagnetic torque in a polyphase induction motor is independent of rotor-circuit resistance.

(*e*) Explain why a single-phase induction motor, as compared to a 3-ϕ induction motor, has larger slip, less efficiency and more noise.

(*f*) Discuss the torque-pulse rate characteristics of a stepping motor.

2. Answer any **two** parts of the following: **(10 × 2 = 20)**

(*a*) A 3-phase induction motor, driving a constant torque load, is connected to constant-frequency voltage source. For this explain the (*i*) with reduction in supply voltage, the stator current increases.

(*b*) A 30 kW, 400 V, 3-phase, 4-pole, 50 Hz induction motor has full-load slip of 5%. If the ratio of standstill reactance to resistance per rotor phase is 4, estimate the plugging torque at full-load speed. Neglect rotational losses and stator impedance. Find also the maximum torque.

(*c*) Discuss the phenomenon of "single phasing" when applied to the three-phase induction motors, designed for continuous working. Explain the difference in behaviour when the single-phasing occurs in the primary and the secondary windings.

3. Answer any **two** parts of the following: **(10 × 2 = 20)**

(*a*) Draw the speed-torque characteristics of single-phase and three-phase induction motors and explain the reasons for which significant differences in their performance.

(*b*) A single-phase, 230 V, 50 Hz, 4-pole induction motor, when connected to ac source gave voltmeter, ammeter and wattmeter readings as 230 V, 2 A and 10 W respectively with the rotor driven at synchronous speed. Neglecting stator leakage impedance and rotational losses, estimate the torque at a slip of 0.05 at rated voltage and frequency. The rotor has a magnetizing reactance of 220 Ω.

(*c*) Explain the slip power recovery scheme of a three-phase induction motor control. What are its advantages and disadvantages?

4. Answer any **two** parts of the following: **(10 × 2 = 20)**

(*a*) Explain the construction and principle of operation of a switched reluctance motor.

(*b*) A 125 kW, 4 pole, 110 V, 50 Hz single phase induction motor delivers rated output at a slip of 6%. The copper loss at full load is 25 watts. Calculate the full load efficiency-and the rotor copper losses caused by the backward field. Rotational losses may be assumed to be 25 watts. Neglect copper loss.

(*c*) Derive expressions for the brush e.m.fs. produced by rotating field in case of a.c. commutator machines. How can the same e.m.f. expressions be obtained by resolving the rotating field into its pulsating field components?

5. Answer any **two** parts of the following: **(10 × 2 = 20)**

(*a*) Discuss the construction, principle of operation and characteristics of universal motors.

(*b*) Discuss the principle of operation, hybrid stepper motors.

(*c*) Write short notes on **TWO** of the following:

(*i*) Linear Induction Motor (LIM)

(*ii*) Repulsion Motors

(*iii*) Shaded-Pole Motors

(*iv*) Double Cage Induction Motor

(*v*) Capacitor Motor.

❑❑❑

Index

P

R

S

T

V

❑❑❑

Win Prizes!

ATTENTION READERS

We request you for your frank assessment regarding some of the aspects of the book given as under:

SPECIAL ELECTRICAL MACHINES

By

SIMMI P. BURMAN

Please fill up the given space in neat capital letters.

(i) What topics of your syllabus that are important from you examination point of view are not covered in the book ?

....

....

....

....

(ii) What are the chapters wherein the treatment of the subject matter is not systematic or organised or updated ?

....

....

....

....

(iii) Have you come across misprints/mistakes/factual inaccuracies in the book? Please specify the chapters and the page numbers.

....

....

....

....

(iv) What other books on the same subject have you found/heard better than the present book ? Please specify in terms of price and quality.

....

....

....

....

(v) Further suggestions and comments for the improvement of the book:

....

....

....

(PS: If need be please attach a separate sheet to write your views.)

Other Details:

(i) Who recommended you the book ? (Please tick in the box near the option relevant to you.)

☐ Teacher ☐ Friends ☐ Bookseller

(ii) Name of the recommending teacher, his designation and address.

....

....

....

(iii) Name and address of the bookseller you purchased the book from :

....

....

....

....

(iv) Name and address of your institution (Please mention the University or Board, as the case may be)

....

....

....

....

(v) Your name and complete postal address :

....

....

....

....

(vi) Course you are enrolled in :

....

....

....

....

The best assessment will be awarded each month. The award will be in the form of our publications as decided by the Editorial Board.

Please mail the filled up coupon at your earliest to:

S.K. Kataria & Sons®

4885/109, Prakash Mahal, Dr. Subhash Bhargav Lane,

Opposite Delhi Medical Association, Daryaganj, New Delhi-110002

Phone: +91-11-43551243; Mobile: +91-9871775858, 7678425209

e-mail: katariabook@gmail.com, katariabooks@yahoo.com

Website: www.skkatariaandsons.com